T0321586

Strategy Games to Enhance
Problem-Solving Ability
in Mathematics

Problem Solving in Mathematics and Beyond

Series Editor: Dr. Alfred S. Posamentier
Chief Liaison for International Academic Affairs
Professor Emeritus of Mathematics Education
CCNY - City University of New York

Long Island University
1 University Plaza -- M101
Brooklyn, New York 11201

Published

Vol. 1   Problem-Solving Strategies in Mathematics:
From Common Approaches to Exemplary Strategies
by Alfred S. Posamentier and Stephen Krulik

Vol. 3   Problems for Metagrobologists: A Collection of Puzzles with Real
Mathematical, Logical or Scientific Content
by David Singmaster

Vol. 4   Mathematics Problem-Solving Challenges for Secondary School
Students and Beyond
by David Linker and Alan Sultan

Vol. 5   Strategy Games to Enhance Problem-Solving Ability in Mathematics
by Alfred S. Posamentier and Stephen Krulik

Problem Solving in
Mathematics and Beyond Volume **05**

Strategy Games to Enhance Problem-Solving Ability in Mathematics

Alfred S. Posamentier
Long Island University, New York, USA

Stephen Krulik
Temple University, Philadelphia, USA

World Scientific

NEW JERSEY · LONDON · SINGAPORE · BEIJING · SHANGHAI · HONG KONG · TAIPEI · CHENNAI · TOKYO

Published by

World Scientific Publishing Co. Pte. Ltd.
5 Toh Tuck Link, Singapore 596224
USA office: 27 Warren Street, Suite 401-402, Hackensack, NJ 07601
UK office: 57 Shelton Street, Covent Garden, London WC2H 9HE

Library of Congress Cataloging-in-Publication Data
Names: Posamentier, Alfred S. | Krulik, Stephen.
Title: Strategy games to enhance problem-solving ability in mathematics /
    Alfred S. Posamentier (Long Island University, USA), Stephen Krulik
    (Temple University, USA).
Description: New Jersey : World Scientific, 2016. | Series: Problem solving
    in mathematics and beyond ; volume 5
Identifiers: LCCN 2016034583 | ISBN 9789813146334 (hc : alk. paper) |
    ISBN 9789813146341 (pbk : alk. paper)
Subjects: LCSH: Games of strategy (Mathematics) | Games in mathematics education. |
    Mathematics--Study and teaching.
Classification: LCC QA270 .P67 2016 | DDC 519.3--dc23
LC record available at https://lccn.loc.gov/2016034583

British Library Cataloguing-in-Publication Data
A catalogue record for this book is available from the British Library.

Desk Editors: V. Vishnu Mohan/Tan Rok Ting

Typeset by Stallion Press
Email: enquiries@stallionpress.com

Printed in Singapore

We dedicate this book of mathematical problem-solving techniques to our future generations so that they will be among the multitude that we hope will learn to love mathematics for its power and beauty!

To our children and grandchildren, whose future is unbounded

Lisa, Daniel, David, Lauren, Max, Samuel, Jack and Charles.
— Alfred S. Posamentier

Nancy, Dan, Jeff, Amy, Amanda, Ian, Sarah, and Emily
— Stephen Krulik

Contents

About the Authors   ix

Chapter 1   An Introduction to Strategy Games   1
Chapter 2   Tic-Tac-Toe Games   9
Chapter 3   Blocking Games   23
Chapter 4   Games Requiring an Ongoing Change of Strategies   37
Chapter 5   Miscellaneous Strategy Games   51
Chapter 6   Suggestions and Strategies for Selected Games   71

Appendix   Game Boards for Various Strategy Games   93

Index   117

About the Authors

Alfred S. Posamentier is currently Chief Liaison for International Academic Affairs at Long Island University, New York. Previously, he was Dean of the School of Education and Professor of Mathematics Education at Mercy College, New York, and before that Distinguished Lecturer at New York City College of Technology of the City University of New York. He is Professor Emeritus of Mathematics Education at The City College of the City University of New York, and former Dean of the School of Education, where he was for 40 years. He is the author and co-author of more than 55 mathematics books for teachers, secondary and elementary school students, and the general readership. Dr. Posamentier is also a frequent commentator in newspapers and journals on topics relating to education.

After completing his B.A. degree in mathematics at Hunter College of the City University of New York, he took a position as a teacher of mathematics at Theodore Roosevelt High School (Bronx, New York), where he focused his attention on improving the students' problem-solving skills and at the same time enriching their instruction far beyond what the traditional textbooks offered. During his six-year tenure there, he also developed the school's first mathematics teams (both at the junior and senior level). He is still involved in working with mathematics teachers and supervisors, nationally and internationally, to help them maximize their effectiveness.

Immediately upon joining the faculty of the City College in 1970 (after having received his master's degree there in 1966), he began to develop inservice courses for secondary school mathematics teachers, including such special areas as recreational mathematics and problem solving in mathematics. As Dean of the City College School of Education for 10 years, his scope of interest in educational issues covered the full gamut educational issues. During his tenure as dean he took the School from the bottom of the New York State rankings to the top with a perfect NCATE accreditation assessment in 2009.

In 1973, Dr. Posamentier received his Ph.D. from Fordham University (New York) in mathematics education and has since extended his reputation in mathematics education to Europe. He has been visiting professor at several European universities in Austria, England, Germany, Czech Republic, and Poland, while at the University of Vienna he was Fulbright Professor (1990).

In 1989 he was awarded an *Honorary Fellow* at the South Bank University (London, England). In recognition of his outstanding teaching, the City College Alumni Association named him *Educator of the Year* in 1994, and in 2009. New York City had the *day*, May 1, 1994, named in his honor by the President of the New York City Council. In 1994, he was also awarded the *Grand Medal of Honor* from the Republic of Austria, and in 1999, upon approval of Parliament, the President of the Republic of Austria awarded him the title of *University Professor of Austria.* In 2003 he was awarded the title of *Ehrenbürger* (Honorary Fellow) of the Vienna University of Technology, and in 2004 was awarded the *Austrian Cross of Honor for Arts and Science, First Class* from the President of the Republic of Austria. In 2005 he was inducted into the *Hunter College Alumni Hall of Fame*, and in 2006 he was awarded the prestigious *Townsend Harris Medal* by the City College Alumni Association. He was inducted into the New York State *Mathematics Educator's Hall of Fame* in 2009, and in 2010 he was awarded the coveted *Christian-Peter-Beuth Prize* in Berlin.

He has taken on numerous important leadership positions in mathematics education locally. He was a member of the New York State Education Commissioner's Blue Ribbon Panel on the Math-A Regents

Exams, and the Commissioner's Mathematics Standards Committee, which redefined the Standards for New York State, and he also served on the New York City schools' Chancellor's Math Advisory Panel.

Dr. Posamentier is a leading commentator on educational issues and continues his long time passion of seeking ways to make mathematics interesting to both teachers, students and the general public — as can be seen from some of his more recent books: *The Circle: A Mathematical Exploration Beyond the Line* (Prometheus, 2016), *Problem-Solving Strategies in Mathematics: From Common Approaches to Exemplary Strategies* (World Scientific, 2015), *Effective Techniques to Motivate Mathematics Instruction* (Routledge, 2016), *Numbers: Their Tales, Types and Treasures* (Prometheus, 2015), *Teaching Secondary Mathematics: Techniques and Enrichment Units,* 9th Ed. (Pearson, 2015), *Mathematical Curiosities: A Treasure Trove of Unexpected Entertainments* (Prometheus, 2014), *Geometry: Its Elements and Structure* (Dover, 2014), *Magnificent Mistakes in Mathematics* (Prometheus Books, 2013), *100 Commonly Asked Questions in Math Class: Answers that Promote Mathematical Understanding, Grades 6–12* (Corwin, 2013), *What Successful Math Teacher Do: Grades 6–12* (Corwin, 2006, 2013), *The Secrets of Triangles: A Mathematical Journey* (Prometheus Books, 2012), *The Glorious Golden Ratio* (Prometheus Books, 2012), *The Art of Motivating Students for Mathematics Instruction* (McGraw-Hill, 2011), *The Pythagorean Theorem: Its Power and Glory* (Prometheus, 2010), *Mathematical Amazements and Surprises: Fascinating Figures and Noteworthy Numbers* (Prometheus, 2009), *Problem Solving in Mathematics: Grades 3–6: Powerful Strategies to Deepen Understanding* (Corwin, 2009), *Problem-Solving Strategies for Efficient and Elegant Solutions, Grades 6–12* (Corwin, 2008), *The Fabulous Fibonacci Numbers* (Prometheus Books, 2007), *Progress in Mathematics*, K-9 textbook series (Sadlier-Oxford, 2006–2009), *What Successful Math Teacher Do: Grades K-5* (Corwin 2007), *Exemplary Practices for Secondary Math Teachers* (ASCD, 2007), *101+ Great Ideas to Introduce Key Concepts in Mathematics* (Corwin, 2006), *p*, *A Biography of the World's Most Mysterious Number* (Prometheus Books, 2004), *Math Wonders: To Inspire Teachers and Students* (ASCD, 2003), and *Math Charmers: Tantalizing Tidbits for the Mind* (Prometheus Books, 2003).

 **Stephen Krulik** is Professor Emeritus of Mathematics Education at Temple University in Philadelphia. While at Temple University Dr. Krulik was responsible for the undergraduate and graduate preparation of mathematics teachers for grades K-12, as well as the inservice alertness training of mathematics teachers at the graduate level. He teaches a wide variety of courses, among them, the History of Mathematics, Methods of Teaching Mathematics, and the Teaching of Problem Solving. This latter course grew out of his interest in problem solving and reasoning in the mathematics classroom. His concern that students understand the beauty and value of problem solving as well as the ability to reason led to his interest in problem solving.

Dr. Krulik received his B.A. degree in mathematics from Brooklyn College of the City University of New York, and his M.A. and Ed.D. in mathematics education from Columbia University's Teachers College. Before coming to Temple University, he taught mathematics in the New York City public schools for 15 years. At Lafayette High School in Brooklyn, he created and implemented several courses designed to prepare students for the SAT examination, while stressing the art of problem solving as opposed to rote memory of algorithms.

Nationally, Dr. Krulik has served as a member of the committee responsible for preparing the *Professional Standards for Teaching Mathematics* of the National Council of Teachers of Mathematics. He was also the editor of the NCTM's 1980 Yearbook *Problem Solving in School Mathematics*. Regionally, he served as president of the Association of Mathematics Teachers of New Jersey, was a member of the editorial team that produced the 1993 publication, *The New Jersey Calculator Handbook*, and was the Editor for their 1997 monograph, *Tomorrow's Lessons*.

His major areas of interest are the teaching of problem solving and reasoning, materials for teaching mathematics, as well as comprehensive assessment in mathematics. He is the author and co-author of more than 30 books for teachers of mathematics, including the *Roads to Reasoning* (grades 1–8) and *Problem Driven Math* (grades 3–8). Dr. Krulik is also the senior problem-solving author for a basal textbook series. Dr. Krulik is a frequent contributor to the professional journals, in mathematics education. He has served as a consultant to, and has conducted many workshops for, school districts throughout the United States and Canada, as well as delivering major presentations in Vienna (Austria), Budapest (Hungary), Adelaide (Australia), and San Juan (Puerto Rico). He is in great demand as a speaker at both national and international professional meetings, where his major focus is on preparing *all* students to reason and problem solve in their mathematics classroom, as well as in life.

In 2007, he was given the Great Teacher Award by Temple University. In 2011 he was presented with the Lifetime Achievement Award for Distinguished Service to Mathematics Education by the National Council of Teachers of Mathematics.

Chapter 1

An Introduction to Strategy Games

One of the most important skills in today's society is problem solving in mathematics and beyond. The ability to resolve problem situations is paramount in today's world. There has been a great deal of research done, which reveals a strong connection between problem solving and playing strategy games. We know that people who are good at problem solving are usually good strategy-game players. Conversely, people who are good at playing strategy games usually show an intuitive approach to successful problem solving. Therefore, throughout this book, we will be presenting a wide variety of strategy games, each of which mirrors in some fashion the kind of thinking that a mathematician engages in during the process of problem solving. It is through this kind of thinking that one enhances the thought process which allows for a strengthened problem-solving capability. Though quite entertaining strategy games are also a training ground for problem solving. Onward now we introduce the process and the various exemplary models of strategy games.

1.1. What is a Strategy Game?

Most people today are used to playing one or more electronic games on devices such as the computer, the i-pad, the smart phone and so on. While these games are indeed enjoyable and do require a certain amount of skill and eye-hand dexterity, many do not require — nor even encourage — careful thinking and strategy development. In some games, time is a major factor. The player is competing with an unknown opponent or against the computer. The two-person strategy games are different. Each game is a head-on, direct competition between two people, each trying to win a game that requires thought and development of a winning strategy. It is true that a person can win some of the time by random play or luck, rather than skill. But this will not happen consistently. After all, it takes time to develop a winning strategy, and when the two players are equally skilled (or unskilled, as the case may be) either player can sometimes win. However, in a two-person strategy game, each player is encouraged to utilize a deeper concentration on developing a needed strategy to succeed at winning the game. Developing a winning strategy should lead to continued success in that game. Finding a strategy that enables one to solve a problem often leads to a general strategy for solving similar problems. In a like manner, as in problem solving, developing and applying a successful strategy at one game may help win at several similar strategy games. Of course, when both players discover the same winning strategy and apply it correctly, many of these games often end in a tie.

1.2. Why Strategy Games?

First of all, we should ask why games? People enjoy playing games. Games are fun! Games are challenging! They provide an opportunity for developing skills and techniques to be successful. People are used to playing games. They expect there to be rules of the game, and a goal for the game. The connection between strategy gaming and mathematical problem solving becomes readily apparent when we examine the heuristics of each, side-by-side. When playing a strategy game, each player must ask oneself some very basic questions, much as is done when engaging in problem solving. Following, we have placed each strategy-gaming heuristic skill directly alongside the corresponding problem-solving heuristic skill to allow for a quick comparison.

Strategy Gaming	**Problem Solving**
1. What are the rules of the game?	1. *Read* the problem
• What constitutes a "win" or a "loss"?	• What's the question?
• What is meant by a "move"?	• What are the facts?
2. What is a good opening move?	2. *Select* an appropriate strategy
• Is there an advantage in going first?	• Apply the strategy
• Should you play defensively?	
3. What strategy can lead to a win?	3. *Solve* the problem
4. Does your strategy work all the time?	4. *Reflect* or *Look back*
• Do you win consistently?	• Does the strategy work on similar problems?

Notice how the strategy-gaming heuristics are similar to those of problem solving. In fact, the heuristics are used over and over throughout the entire game. Each time a player makes a move in a strategy game, that player presents the opponent with a new problem. Players constantly ask themselves, "How do I counter that move with one of my own?" This is all part of developing a game strategy.

1.3. Finding and Creating Strategy Games

There are many books that contain one- or two-person strategy games. Some of these games will be discussed in this book. However, you can often create some of your own games using either or both of the following techniques:

• Change the goal of the game.
• Change the shape of the given playing surface, game pieces, or game board.

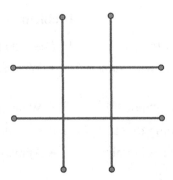

Figure 1.1.   Tic-tac-toe board.

Let's take an example of each. Most people are familiar with the basic game of tic-tac-toe (sometimes called "Noughts and Crosses" in Great Britain). About 500 BCE, Confucius describes a game called *Yih*, which is today's tic-tac-toe. Yet in the Western world the game seems to have its origins in ancient Rome and Egypt and has come to its current form in the nineteenth century. In this game, players alternate turns placing an "X" or an "O" in an empty cell on a 3 × 3 or 9-cell grid (see Figure 1.1).

The goal is to get three of your own marks in a straight row, with no intervening spaces or opponent's mark. Most of us have played this game and developed a strategy that will lead to either a win or a draw. The majority of players usually prefer to go first. Since there are nine cells on the board, going first ensures a chance to place your mark in five cells to the opponent's four cells. Furthermore, most players going first place their mark on the center square. This square is involved in a winning position 4 times out of 8. Corner squares are involved in a winning position 3 times out of 8. Sounds easy enough, doesn't it? And yet, the center cell as a first move is not the best approach. The player going first should place their X or O in one of the four corner cells. This corner-cell placement is the one that can lead to developing two possibilities for a win. Once these two possibilities have been developed, the opponent cannot block both at once, and a win is assured. If both players are careful and know the strategy, then the very least that can occur is a tie.

Let's inspect the strategy used in playing tic-tac-toe. The player to go first, using the X, has three options for his first move. He can either place the X in a corner cell, a middle border cell, or in the center cell. This player can force a draw or can win with any of these three starting moves. However, by placing the first move in a corner cell, the second player is limited in his move to avoid losing.

The second player, using an O, must respond to the first player's move defensively. If the first player chose to place the X in a corner cell, then the second player must place the O in the center cell. If the first player chose to place the X as the initial move in the center cell, then the second player must place his O in a corner cell. If the first player places his initial move by placing the X in a side middle cell, then the second player must place the O either at the center cell, or at a corner cell next to the O-placed cell, or at the side middle cell opposite the X cell. Any other move by the second player will result in a win by the X player. After these two moves, the second player continuously places Os to block the first player's attempt to get three Xs in a row. Of course, there is always the possibility that the first player, X, might make a careless move allowing the second player, O, to win rather than just to force a draw.

For the second player, placing O, to guarantee a draw, he would need to adhere to the following: If the first player does not place the X in the center cell, then the second player should occupy the center cell with the O, followed by a side middle cell. If the first player places an X in the corner square, and the second player places and O the opposite corner, that will allow the first player to win if he places the second X in one of the occupied corners. If the first player places an X in the corner cell, then the only way the second player in force a tie to place an O in the center cell and then, as his next move he would place an O in a side middle cell. By this time you should be able to notice the various strategies that clever players would use to always end up in a tie. It is when one player makes a careless move the other player will be able to win.

However, let's make a simple change in the game, thus creating another game and requiring another strategy. In this new version of the game, three-in-a-row *loses* the game. Does the strategy change? Do you still want to go first? Do you still want to place your X or O in the corner square? A new problem has been presented, and a different strategy must be developed. Play the game with this new goal and see what transpires.

A second approach to creating a different game is to make a change to the game board. In a new game of tic-tac-toe, we might play on a 5 × 5-cell square board, but require 4 in a row to win. Now does the strategy change since there are 25 cells? What's the strategy here? Or, perhaps, change the board to a 4 × 4-cell board of 16 cells, and try to get three-in-a-row. Now does it make sense to go first? After all, each player will have a chance to fill eight cells. Should we still play 3 in a row wins the game? Or 4 in a row? The strategy for winning in each version of the game has changed! We will consider these variations later in the book.

Even a strategy game as simple as original tic-tac-toe affords many opportunities to utilize mathematical skills. Explaining the rationale for making a particular move helps develop reasoning skills. The communicating of these ideas verbally, creates an informal dialogue, which leads to increasing the higher order thinking and problem-solving skills. Every move requires an informal calculation of the probability of winning or losing the game. Even an opponent's move raises the question of the probability of winning the game.

This book is divided into chapters, each considering a different type of strategy game. Sometimes a single strategy can resolve several games, similar to applying a specific problem-solving strategy to several different problems. Other times each game may require a specific strategy. In some cases, we might provide a hint or a suggestion. We will not, however, reveal a complete strategy for a particular game until the last chapter of the book (Chapter 6), so that you can have time to craft your strategy and then if you wish, compare it to our strategy in the last chapter. We don't want take away the fun of you, the reader, trying to develop your own strategy. Where a specific type of board is given, feel free to copy the board, enlarge it, and play the game as many times as it takes to discover a winning strategy.

Many of the games in this book have their origins in countries around the globe. Some have been played for centuries by children and adults alike. The players use game boards drawn on the sand or dirt, or painted on leather animal hides. Rocks or pebbles of different shapes or colors make up the playing pieces. Some games are available commercially with more permanent game boards made from wood, plastic or metal.

Most of the games in this book do not require any special equipment. Generally, all you need is a pencil, paper, a copy of the particular game board (most game boards are available in the Appendix following Chapter 6), and possibly some toothpicks, chips or coins, which should be easily identifiable for each player. We wish you lots of enjoyment.

Chapter 2

Tic-Tac-Toe Games

In this chapter, we will consider some variations of tic-tac-toe games. Most people are familiar with the basic game, which we described in the previous chapter. This is the one played on a 9-cell (3 × 3) square grid. Players alternate turns placing one of their markers (usually an X or an O) in any empty cell. As we discussed in Chapter 1, the objective is to be the first player to get three of their own marks in a row, horizontally, vertically or diagonally. Players usually can quickly develop a winning strategy, although if both players understand the game, a tie is most often the result. Let us now move along to other variations of this time-honored game of strategy.

2.1. Losers Tic-Tac-Toe

As we indicated in Chapter 1, a simple change in the rules and/or goal of the game causes some changes in strategy formation. Here we will consider a player who gets three of his marks in a row loses the game.

Rules of the Game

The game is played on the traditional 3 × 3 (or 9-cell) grid. As in regular tic-tac-toe, players alternate turns, placing one of their marks (X or O) in any empty cell anywhere on the grid.

The Goal of the Game

The goal of the game is NOT to get three of a player's marks in a row. Three in a row loses! This is clearly the opposite goal of the traditional version of the game.

Sample Simulation of the Game

Suppose the first player places an X in the center cell. If the second player places an O in a corner cell, then the first player is forced to have two of his marks in a row in more than one position. How can this be avoided? Was this a good first move? Notice that once a player occupies the center cell, their next move *guarantees* two in a row, which turns out to be a bad strategy indeed. We encourage you to try your skill at this game and then read the strategy for winning in Chapter 6.

2.2. Triangular Tic-Tac-Toe

For this variation of tic-tac-toe, we will change the shape of the game board into a triangular shape, as shown in Figure 2.1.

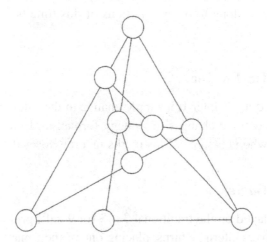

Figure 2.1.   Triangular tic-tac-toe board.

Rules of the Game

Once again, players alternate turns placing one of their markers (X or O) in any empty circular cell.

The Goal of the Game

The object of this game is the same as in regular tic-tac-toe; that is to get three of your markers in a row to win the game.

Sample Simulation of the Game

The first player begins by placing an X in one of the circular cells, perhaps one of the three center cells. Then the second player puts a mark, O, in one of the corner cells, which are at a vertex of the triangular board. Now the first player puts his or her mark, X, in a cell of the inner triangle.

Does player X have a guaranteed win? Why or why not? Were these moves good ones to begin with? Is the strategy similar to that of regular tic-tac-toe? Does the shape of the board cause some confusion at first? Once again we urge you to play the game a few times and then compare your strategy to the one suggested in Chapter 6.

2.3. Achi

The game of achi was found carved into a roof slab of a temple in Al-Qurna, Egypt, and dates back over 3500 years. It is still played around the world today, and is particularly popular in Ghana.

Rules of the Game

The game is played **on** the traditional 3 × 3 (or 9) square-cell grid. However, the circular cells are joined by lines as shown in Figure 2.2. Each player has three chips or markers that are easily identifiable. Players alternate turns, placing one of their markers into any empty cell anywhere on the board. If all six markers have been placed and no person has won,

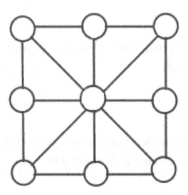

Figure 2.2.   An achi game board.

players continue alternating turns sliding any one of their own markers along one of the straight lines into an empty adjacent cell.

The Goal of the Game

The goal of the game is to be the first player to get three of his or her own markers in a straight line, horizontally, vertically or diagonally.

(*Note*: In some versions of achi, the game can also be played with each player getting four chips, or markers, instead of three. Otherwise, the rules remain the same.)

Sample Simulation of the Game

As in most games of tic-tac-toe, the opening move is critical. Player A places a marker in any of the border cells. After the second player places a marker, Player A tries to get a second marker so as to provide two opportunities to win. Place the chip adjacent to the first one. This will force Player B to play defensively and not concentrate on winning. Can the game end in a tie? Does going second force the player to play defensively?

2.4. Four-in-a-Row

Rules of the Games

The game board is usually a 7 × 6 array of square cells as shown in Figure 2.3. Players alternate turns placing one of their own marks (chips

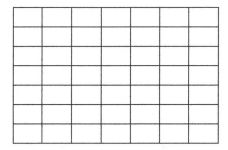

Figure 2.3.   A four-in-a-row game board.

of different colors can also be used, such as red and black checkers) anywhere in an empty cell, on the board, with one restriction: Each marker placed must be in the lowest empty cell in the column where it is to be placed.

The Goal of the Game

To win the game, a player must be the first to connect four markers in a row, horizontally, vertically or diagonally.

Sample Simulation of the Game

To follow this strategy, look at Figure 2.4. Suppose Player 1 puts a mark (X) in cell number one. Then Player 2 might go in the second box to the right with an O. This blocks the first player from a horizontal win. Player 1 puts the next mark (X) above the first O. Player 2 places an O to the right of his previous O. Player 1 places an X above the second O. Player 2 continues to try for four in a row in the bottom row and places his third O alongside the first two Os. Player 1 must now block this attempt with an X in the first row. Player 2 places an O above the third O to block the Player 1 from making a row of four in the second row. Player 1 now places an X in the third row and third column. Now see what would happen if Player 2 were to place an O in the cell marked "?" in an effort to get three Os in a vertical row. Player 1 would then be able to place the next X above this third O in the fourth column and win the game with a diagonal line of four Xs. It is best to play this a bit to see further strategies.

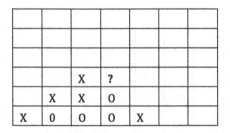

Figure 2.4.   A sequence of moves in four-in-a-row.

2.5. Pyramid Tic-Tac-Toe

Here is another version of tic-tac-toe but played once again on a differently shaped game board.

Rules of the Game

The game follows the same rules as in regular tic-tac-toe. That is, players alternate putting one of their marks (X or O) in any empty cell. There are seven ways to win: four horizontally, two diagonally, and one vertically.

The Goal of the Game

The goal of the game is to get three of one's own marks consecutively in a row with no spaces nor opponent's marks in between. The row can be made diagonally, horizontally, or vertically (see Figure 2.5).

Sample Simulation of the Game

Notice that there are seven ways to get three of one's marks in a row. Player A puts an X-mark in the lower left corner. Player B puts an O-mark in the top cell. Player A now puts an X in the second cell in the bottom creating a potential win. Can Player B block and create a potential win at the same time? Can either player guarantee a win with their next move? (See Figure 2.6.)

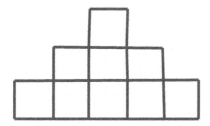

Figure 2.5. Game board for pyramid tic-tac-toe.

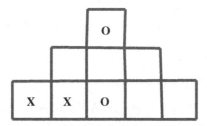

Figure 2.6. A game of pyramid tic-tac-toe.

2.6. Tac-Tix

Rules of the Game

The game is played on a 4 × 4 array of cells. Each player uses four chips, coins or other identifiable pieces. The starting position is as shown in Figure 2.7. Players alternate turns moving one of their own pieces to any adjoining vacant cell. They may move vertically, horizontally, forwards or backwards, but never diagonally. No pieces can be captured, jumped or removed from the board.

The Goal of the Game

The goal of the game is to be the first person to get three of his own pieces in a row, vertically, horizontally, or diagonally with no intervening empty, or an opponent's occupied, cells in between.

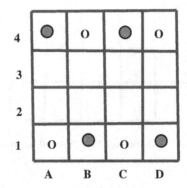

Figure 2.7.   Starting position for the game of tac-tix.

Sample Simulation of the Game

Suppose the Player 1 moves from cell A-4 into A-3. Player 2 moves from C-1 to C-2. Player 1 might then move into now-vacant C-1, thus beginning to establish a diagonal line. This forces Player 2 to begin to defend rather than play to win. Can layer 2 win? Was there a better first move for Player 2? Continue this game to see where it goes.

2.7. Put them Down Tic-Tac-Toe

Rules of the Game

Each player has three markers or chips that are easily identifiable. Players alternate turns placing one of their chips in any empty circle on the game board. The center circle cannot be used by either player as their first move (see Figure 2.8).

Once all six chips are placed, each player in turn, moves one of his or her own piece along any line into an adjoining vacant space, anywhere on the board.

The Goal of the Game

The first player to get three of his or her own pieces in a row (horizontally, vertically, or diagonally) with no spaces in between is the winner.

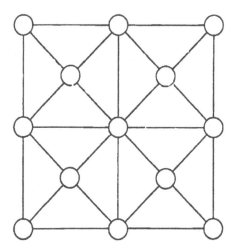

Figure 2.8. Put them down tic-tac-toe board.

Sample Simulation of the Game

We will denote the players with the odd and even numbers. Suppose Player 1 (using odd number chips) puts a chip in the upper left circle. (Is this a good move? Why? Or why not?) Player 2 places a chip (even numbered) in the upper center cell. Now the center cell may be used, so Player 1 now selects the center cell. This forces Player 2 to block with "4" (see Figure 2.9). Player 1 then places the "5" as shown in Figure 2.9.

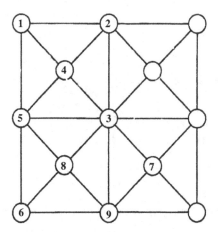

Figure 2.9. The start of a sample game.

Player 2 doesn't have to worry about a horizontal (middle) win by Player 1, since he would have to slide a chip there, and cannot do it from the current stand, so he places the last chip "6" at the lower left.

Is a win now guaranteed? Which player can win? Can a tie be forced? Play out the game as shown. The "3" is moved to the "7" position, and the "6" is moved to the "8" position, and so on.

2.8. Circle Tic-Tac-Toe

Like most versions of this old-time game, changing the shape of the board creates a whole new series of moves and strategies. The circular shape often causes some confusion at first.

Rules of the Game

Circle tic-tac-toe is played on a polar coordinate type graph grid as shown in Figure 2.10.

The board is made up of four concentric circles of radius 1, 2, 3 and 4, as well as twelve radial line segments placed every 30°. Notice

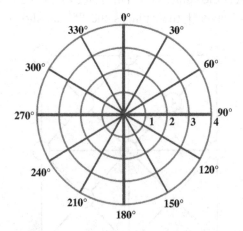

Figure 2.10. A game board for circle tic-tac-toe.

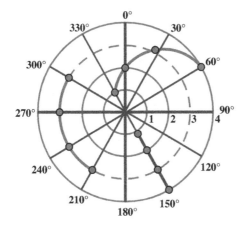

Figure 2.11. Winning positions for circle tic-tac-toe.

that the origin has been omitted since that point would yield a distinct advantage to the player who goes first. Players alternate turns placing their mark (X or O) on an intersection of a circle and radial line segment.

The Goal of the Game

The winner is the first player who gets four of his or her marks in a consecutive row, either along a radial line segment, along a circle, or in a spiral (see Figure 2.11).

Sample Simulation of the Game

Player A puts an X on the outer circle (circle 4) at the point (4, 270°). Player B counters with a mark at (2, 90°). Player A now goes to (3300°). After each player has made three moves, the board looks like Figure 2.12.

It is Player A's turn. Can he win? How? What if it were player B's turn? Should Player B block ? Why? Play it out to see how this strategy can develop.

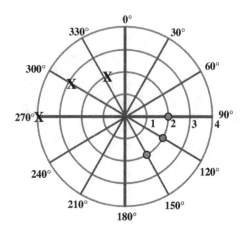

Figure 2.12.   Sample circle tic-tac-toe game.

2.9. Circle Tic-Tac-Toe (Simplified)

If the previous game sounds complicated at first, you might begin by play-ing a simpler version.

Rules of the Game

The game board here utilizes only three circles, radius 1, 2, and 3, as shown in Figure 2.13. The rules remain the same and the angle measures remain the same. Players again alternate turns placing a mark on any intersection of a radial line segment and circle.

The Goal of the Game

Again, the players alternate turns in an attempt to get *three* in-a-row either along a circle, a radial segment, or in a spiral.

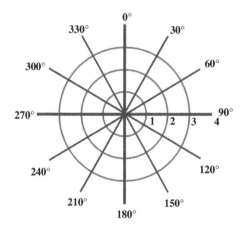

Figure 2.13.   A simpler circle tic-tac-toe game board.

Sample Simulation of the Game

Same procedure as the previous version. Once again, we urge you to try the game after having read the sample simulation, and only after you have a reasonably good command of the game. Read the strategy that we provide in Chapter 6.

Chapter 3

Blocking Games

In some strategy games, players concentrate on not only winning the game by developing a particular strategy, but also setting up a defensive strategy to prevent the opponent from winning. In the latter case, they must focus on stopping or blocking their opponent from developing a similar winning strategy. In this chapter, we will consider a few such games, where a defense could be as useful as a winning strategy.

3.1. Nim

The game of nim has appeared in many places, using different numbers of chips in each version. The most common approach uses 32 chips or tokens as presented here.

Rules of the Game

Begin by placing 32 chips (or coins) on the table between the two players. Alternating turns, each player in turn can remove one, two, or three chips from the pile of chips.

The Goal of the Game

The winner is the player who picks up the last chip (or coin) from the pile.

Sample Simulation of the Game

Suppose Player A goes first and selects one chip, leaving 31 chips. Player B also removes one chip leaving 30 chips. Player A now takes two chips, leaving 28 chips. Player B makes the same move, removing two chips and leaving 26 chips. Player A now removes three chips leaving 23 chips. Player B follows, making the same move and leaving 20 chips. If play continues in this same way, who will win? Can either player change the pattern and win? Was selecting one chip a good starting move? Why or why not?

3.2. Reverse Nim

Once again, we create a different game by changing the ultimate goal of the nim game presented in Section 3.1. A different game requires a different strategy as we shall see.

Rules of the Game

The rules for reverse nim are the same as for nim; that is, players alternate turns removing one, two, or three chips from the pile of 32.

The Goal of the Game

The goal of this game is NOT to pick up the last chip, rather to force the opponent to do so. Thus, the person who picks up the final chip is the loser.

Sample Simulation of the Game

The game might begin in the same manner as in the regular game of nim. That is, Player A takes one chip as does Player B. Player A now takes two chips. Player B does the same. Will the sequence of moves change the winning player? Was this a good start? How many chips should be left at the end in order to assure a win?

3.3. Domino Cover Up

Most people are familiar with a domino, which is composed of two squares joined together forming a rectangle. This game requires a traditional 8 × 8 checkerboard, and a set of 32 dominoes, each of which covers exactly two squares – horizontally or vertically but not diagonally – of the checkerboard.

Rules of the Game

Players alternate turns placing a domino either horizontally or vertically on the board, and thereby covering exactly two squares.

The Goal of the Game

A player wins when he/she successfully places a domino on the board and the opponent has no possible way to place a domino on the board.

Sample Simulation of the Game

Player A decides to cover squares in the top row of the board and places a domino in a horizontal position to cover two boxes in the row. Player B decides to place his or her dominoes in a vertical move each time, but randomly selects to cover any two boxes. Each player continues by following their plan. Which player will win? Is there a better first move? Can Player A block Player B as well as setting up a final successful move?

3.4. Tromino Cover Up

A tromino is a geometric figure made up of three congruent squares joined together along an edge as shown in Figure 3.1. These are the only two shapes for trominos. The game requires that eight of each tromino be available to start. Each should cover exactly three squares.

Figure 3.1. Tromino shapes.

The game is played on a 5 × 5 square grid, as shown in Figure 3.2

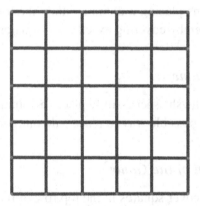

Figure 3.2. Tromino game board.

Rules of the Game

Players alternate turns, placing a tromino of either shape anywhere on the game board, covering exactly three squares.

The Goal of the Game

The goal of the game is to be the player who places the last tromino successfully on the game board, and thus, at the same time, blocks the opponent from making a move – that is, by leaving the opponent no three squares on which to place the trominos.

Sample Simulation of the Game

Let's suppose the first player places an "L-shaped" tromino in the center of the game board. The second player then places a "line-shaped" tromino at the top of the board, horizontally covering three boxes. The first player then places another "L-shaped" piece so as to isolate a single box. Who will win if play continues in this way? Is there a better move? Can either player guarantee a win by going first? By going second? Which player has made a better first move? Suppose we change the goal of the game so that the player who places the last tromino loses, would the strategy change?

3.5. Pathways

Pathways is a two-person game that forces you to plan ahead. You must not only concentrate on your own moves, but also try to block your opponent's moves at the same time. The game is played on a square array of dots as shown in Figure 3.3.

Rules of the Game

Players alternate turns. Player A draws a line segment connecting any two of the dots. The dots must be next to each other, and the line segment that

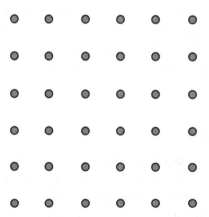

Figure 3.3. Pathways game board.

joins them can connect them either horizontally or vertically, but not diagonally. Player B then connects any two dots that are not already connected. Again, the dots must be adjacent and the segment must be horizontal or vertical. Neither player can cross an opponent's path, which is defined as any two collinear segments.

The Goal of the Game

The goal of the game is to be the player who first makes a continuous trail of segments going from one side of the board to the other – vertically or horizontally.

Sample Simulation of the Game

Player A connects two dots at the top of the board. Player B makes the same move at the other end. If Player B continually copies Player A's move, who will win? Is it better to connect dots at random? Or concentrate on blocking your opponent?

3.6. Hex – Board Game

Hex is a game that has been played for many years. It was invented by the Danish mathematician Piet Hein in 1942 and further produced by the American mathematician and Nobel Laureate, John Nash in 1947. The game has been played on a variety of board sizes, but most traditionally is played on a 11×11 array of hexagons, as shown in Figure 3.4.

Rules of the Game

One player attempts to make a line horizontally (from side to side) marking X's in his or her cells, while the other player attempts to make a line to be from top to bottom (vertically) marking O's in his or her cells. Players alternate turns placing one of their marks in any open cell anywhere on the game board. Notice that the four corner hexagons can be used by either player since they occur on two adjacent sides.

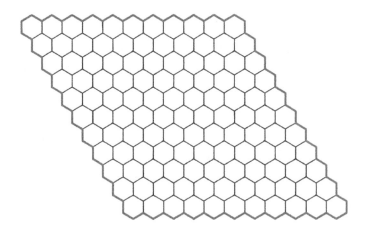

Figure 3.4.   Hex game board.

The Goal of the Game

The goal of the game is to form an unbroken line of the player's marks from one side of the board to the other, while preventing the opponent from doing the same with the other two sides. The winner is the player who develops an unbroken path across the board from side to side. John Nash proved that this game cannot end in a tie.

Sample Simulation of the Game

Player A takes one of the corner cells to put their mark. Player B does the same at the opposite corner of the board. Whenever Player A makes a move, Player B copies the move on the other end of the board. Player A now selects the next box that is immediately adjacent to his or her original move, and begins to make a straight line across the edge of the board. Who will win? Should either player change strategy? Why? Is it better to block or attempt to win?

3.7. Sprouts

The game of sprouts was invented by two British mathematicians John H. Conway and Michael S. Paterson in the early 1960s. It requires little equipment – just pencil and paper.

Rules of the Game

Begin by drawing three "spots" on a piece of paper (see Figure 3.5).
     There are two kinds of moves in the game. A player can make either of the moves in his or her turn.

Figure 3.5.   Starting a game of sprouts.

(1) Connect any two free spots on the board. Place a new spot in the middle of the newly drawn line.
(2) Connect any free spot to itself, and place a new dot in the middle of the newly drawn line.

     At the same time, there are two additional rules that must be followed:

(3) No spot may have more than three lines either leading into or away from the spot (see Figure 3.6).
(4) While lines may be curved, no line can cross another one.

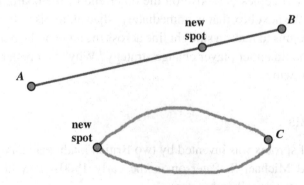

Figure 3.6.   Two possible moves in sprouts.

Notice that in each case shown, the new spot has two lines drawn to it, and is still open, but for only one more line to touch it. Then it would have three lines to it, which is the most it could have.

The Goal of the Game

The player who cannot make a legitimate move is the loser.

Sample Simulation of the Game

Let's look at a typical game of sprouts. We shall start with three points as shown in Figure 3.7a.

Player number one connects A to B and adds a new point, D (see Figure 3.7b).

Player number two connects C to C (itself) and adds a new point E as shown in Figure 3.7c.

Player number one now connects D to E and adds point F, as we have done in Figure 3.7d.

Notice that both points D and E are now considered "dead"; neither can be used again. Each has three lines emanating from it (see Figure 3.7d).

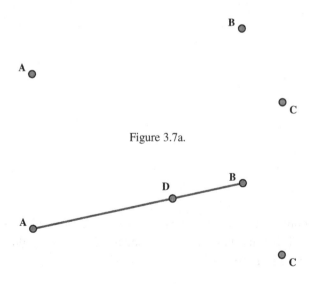

Figure 3.7a.

Figure 3.7b.

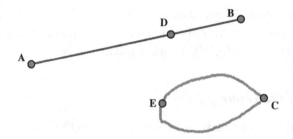

Figure 3.7c.

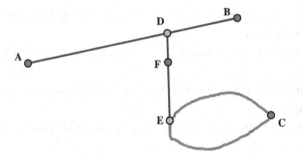

Figure 3.7d.

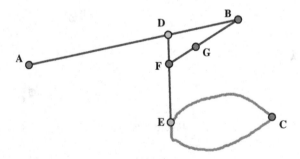

Figure 3.7e.

Player number two now connects B to F and adds G as shown in Figure 3.7e. Player number one now connects G to A, and adds point H, as we have done in Figure 3.7f.

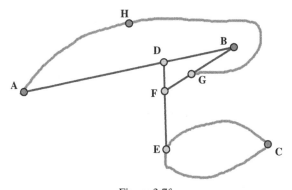

Figure 3.7f.

Play continues in a similar manner until one player cannot make a valid move. Was player number two playing a defensive game? Or was the player trying to win?

3.8. Line-Up-Nim

Line-up-nim is a variation of the game of nim that was described in Section 3.1.

Rules of the Game

Place 15 matchsticks in a five-line array as shown in Figure 3.8. The game can also be played by simply drawing five rows of lines as shown in Figure 3.8.

Rules of the Game

Players alternate turns crossing out one, two, or three of the lines (or removing one, two or three matchsticks) from the array. There are two restrictions, however. First, the lines crossed out (or matchsticks removed) must all come from the same row. Second, those removed must be adjacent to one another with no line remaining between.

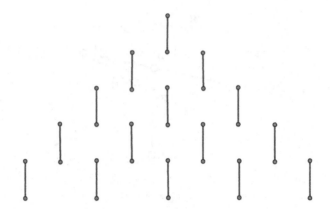

Figure 3.8.   A starting position for a game of line-up-nim.

The Goal of the Game

The objective of the game is to force your opponent to remove the last matchstick or cross out the last line. The player who takes the last matchstick is the loser.

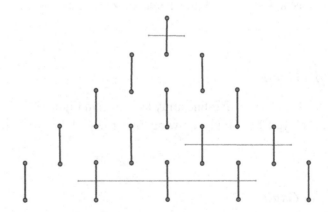

Figure 3.9.   Sample moves in a game of line-up-nim.

Sample Simulation of the Game

Player A removes three from the bottom row. Player B now removes one from the top row. Player A then takes two from the third row. This leaves the game board as shown in Figure 3.9.

What move should Player B now make? Can Player B remove the two remaining lines in the bottom row? Why, or why not?

Bear in mind that the game can be played with six, seven or even eight rows continuing the arrangement pattern as shown. Obviously, the strategy to win changes as the array changes and the number of lines as well.

3.9. Loop-de-Loop

This game is actually similar to sprouts (see Game 3.7) but a much simpler game. It rarely ends in a tie.

Rules of the Game

Start with an arrangements of 12 dots as shown in Figure 3.10. Other arrangements of dots with fewer, or more than 12, can also be used. However, we find 12 to be a good number to begin with when first starting to learn the game.

Players alternate turns drawing a loop through one, two, or three dots. The following restrictions must be met:

* Loops cannot cross one another.
* A new loop may not pass through a dot which is included in any other loop.

Figure 3.10. A typical array of dots for the game of loop-de-loop.

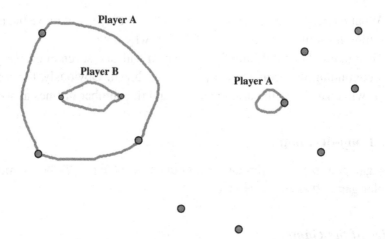

Figure 3.11.   The first three moves in a simulated game.

The Goal of the Game

The player who makes the last possible loop is the winner.

Sample Simulation of the Game

Suppose Player A begins by making a loop that connects three dots in the left side array. Player B now decides to make a loop through the two dots remaining inside the first loop. Notice that the left side of the board is now "dead". That is, no possible moves remain on this part of the board.

Player A now connects one dot to itself on the other side of the board (see Figure 3.11).

Is this a winning position for Player A? Player B? Why?

Chapter 4

Games Requiring an Ongoing Change of Strategies

Many games require an ongoing change of strategy as the game progresses. Sometimes, you play a defensive-type strategy to keep your opponent from winning. Other times this can change to moves that display an offensive-type strategy. Sometimes, this change of approach "just happens" without a player realizing it. In this chapter, we will consider some games that invoke such fluctuating strategies.

4.1. Nine Men's Morris

The game, which dates back to the Roman Empire, has each player holding nine chips, or tokens, and each set of nine is of a different color. These should be readily identifiable as to which player is moving which color. In addition, a copy of the game board is needed (see Figure 4.1). The game is sometimes referred to as Nine-Man-Mill.

Rules of the Game

Players alternate turns placing one of their nine chips on any empty cell on the board. A player tries to get three of his or her own tokens in a row, horizontally or vertically, all the while trying to avoid having the opponent

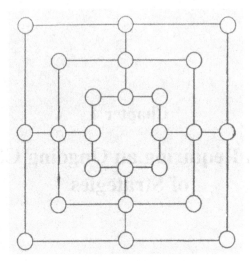

Figure 4.1.   Game board for Nine-Man Morris.

block this attempt. Upon getting three of his or her own chips in a consecutive row, this player can then select any one of the opponent's chips to be removed from the board.

Once all 18 chips have been placed on the board, players take turns moving their chips to any adjacent non-occupied cell. Every time you form an uninterrupted row of three chips in a straight line, you get to remove any one of your opponent's chips.

The Goal of the Game

The goal of the game is to reduce your opponent's chips to fewer than three. Thus, they can no longer get three in a row and so the game is over.

Sample Simulation of the Game

Player A puts a chip in a corner cell. Player B immediately takes the horizontally adjacent cell to block a potential horizontal move. Player A places a chip next to his first chip, attempting to create a vertical row of three chips. Player B must block Player A from forming a vertical line of three chips, or else player B will lose one of his chips. Does this suggest that going first has an advantage? Should Player B stay on the defensive? Why? Is a corner cell a good initial move?

4.2. Hare and Hounds

Rules of the Game

The game, which dates back to medieval northern Europe, is played on a standard 8 × 8 checkerboard. One player has three chips (the hounds); the other player has one chip of a different color (the hare). Only the dark colored squares are used as in the game of checkers. To start the game, the three hounds are each placed on the dark squares at one edge of the board. The hare is placed on any one of the four dark squares at the opposite end-line of the board.

The players alternate turns moving to one dark square (diagonally) at a time. The player controlling the hounds can move any of his or her pieces one at a time in turn. The hounds can only be moved forward, never backwards. The hare can move one square (diagonally) at a time, but can move forward or backward. There is no jumping.

The Goal of the Game

The hare wins if it can move into any of the original squares on the side of the board from which the hounds began. The hounds win if they "box" the hare in so he cannot move.

Sample Simulation of the Game

Player A places his "hare" in either end box of the back row. Player A moves his piece forward one square. Player B moves a piece (the hounds) from the opposite side towards the center. Player A now moves back and forth from this square to the original square. Player B moves another piece towards the center of the board, trying to keep his "hounds" in a straight line. Will Player A be able to get the "hare" *behind* the hounds and thus win? Or will the hounds trap the hare so he cannot move? Was the hare's first move a good one? Why? Is the back-and-forth strategy a good one? Is starting in an end box a good move?

4.3. Kalah

This game was invented in 1940 in the United States by William Julius Champion, Jr. (1880–1972) and has maintained its popularity. Commercially manufactured versions of this game board are readily available. However, an egg carton with two rows of six openings is easy to make. These are referred to as "pits". Place two saucers, one at each end. These are the "home" or "house" for each player, the home to their right. In addition, the game requires 36 pebbles, lima beans or coins. These are referred to as "stones".

Rules of the Game

The rules for the game will probably sound a bit complicated at first. But, after you have played the game a few times, they should become clear. Set up the game board as shown in Figure 4.2. There are three pebbles – called "stones" – in each of the 12 "pits" as shown. One player sits in the North seat, the other sits opposite in the South seat. All moves are made counter-clockwise.

The players alternate taking turns. The player whose turn it is to move, takes all the stones from any one of his or her own pits. The player then moves counter-clockwise, distributing one stone in each pit including the player's own house. The player continues distributing the stones in this manner, even including the opponent's pits, but not the opponent's house (which is skipped).

NORTH

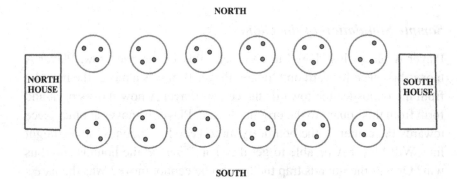

Figure 4.2.   Kalah game board in starting position.

- If the last stone is dropped in the player's house, he/she takes another turn.
- If the last stone ends in a pit on the opponent's side or in his or her own side, then his or her turn is over.
- If the final stone being distributed ends in an empty pit on the player's own side, the player takes all the stones in his or her opponent's pit directly opposite this pit and deposits them in his or her own house. Their turn is then over.

The game ends when one player has no stones in any of his or her pits, and thus cannot move.

The Goal of the Game

At the end of the game, the player with the most stones in his or her "house" is the winner.

Sample Simulation of the Game

Player A takes the three stones from his third pit and drops one of them in pit number 2, one in pit number1, and the third one in his or her home. The player goes again, since he ended in his home. He/she now decides to take the four stones in pit number 2, and drop one in pit number 1, one in his home, and one in each of his opponent's two pits. The player's turn is over. Player B makes a similar first move, taking the stones from his or her own pit number 3 and distributing them into his or her own pit number 2, pit number 1 and home. Player B takes another turn.

Play continues in this manner until one player empties the stones from all his or her pits. The players now count the number of stones in their respective home to determine the winner, who is the one who has more stones in his or her house.

4.4. Split-'Em-Up

Split-'em-up is another game involving two players. It involves a great number of blocking strategies throughout.

Rules of the Game

The game requires 32 chips (or matchsticks, or toothpicks, etc.) placed between the two players. Each player takes a turn separating the pile of chips into two piles of *unequal* size. For example, if a pile of 6 chips is to be split, it can be separated into two piles of 5 and 1, or into two piles of 4 and 2, but it *cannot* be separated into two piles, each containing 3 chips, since the rules of the game require that the splits be into unequal sized piles. Thus, a pile of 2 chips can never be separated, since the only separation would be 1 and 1, which is not allowed.

The Goal of the Game

The first player who cannot make a successful separation, or split, loses the game.

Sample Simulation of the Game

Player A starts by dividing the 32 chips into piles of 2 and 30. Player B splits the pile of 30 into 16 and 14. Player A divides the 16 into 14 and 2. There are now only two piles of 14 each that can be split. The piles of two can never be touched again, since they could only be split into two equal piles, both containing 1 chip, which is not allowed. Player B splits one pile of 14 chips into 6 and 8. What would be a good move for Player A to split now? Why? Is splitting so one pile contains two chips always a good move?

4.5. OZO

The game of OZO is usually a two-player game, but it can be played by three players as well. It is similar to tic-tac-toe, but it is much more complex, since players need not make a specific mark when it is their turn.

Rules of the Game

The game is played on an eight-by-eight square game board. Players alternate turns placing a Z or an O in any empty cell, anywhere on the board.

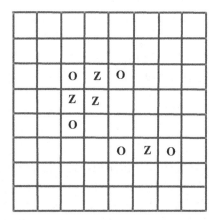

Figure 4.3.   Some scoring positions for OZO.

Players can place either mark – they need not use the same mark when it is their turn. The players attempt to make a three-cell sequence O–Z–O in a straight line, horizontally, vertically or diagonally (see Figure 4.3 for some examples).

If a player completes an O–Z–O line, that player's turn continues. Otherwise, it becomes the other player's turn. A player scores one point for each O–Z–O line he or she makes.

The Goal of the Game

When the board is full and/or no more sequences of O–Z–O can be made, the game ends. The player who has the most points is the winner.

Sample Simulation of the Game

To start the game, Player A puts an O mark in a corner square of the board. Player B does the same at the opposite end of the game board. Neither player wants to put a Z near this first move, since the opponent would immediately complete a sequence of O-Z-O, and then score a point. Instead, Player A puts another O next to his original O mark. Player B decides to mark a Z towards the middle of the board. The game continues with players putting either an O or a Z in any vacant square anywhere on the board.

Was the first move a good one? How many possible ways can O-Z-O be made from this starting position? Was Player B's second move a good one? Why, or why not?

4.6. Lau Kata Kati

This is a game from India, similar in many ways to pretwa (see Figure 4.5), but with a differently shaped game board and a bit easier. Like many of these games, the pieces are all placed on the board at the start, with only the intersection at the center vacant. This forces the first move. You might wish to learn how to play lau kata kati before attempting pretwa.

Rules of the Game

The game is played on a game board consisting of two congruent triangular grids, meeting at the center (see Figure 4.4). Lines are drawn horizontally on each side of the board, and an additional line is drawn from top to bottom through the common vertex at the center. This gives a total of 19 intersections. The chip pieces are shown in the starting position.

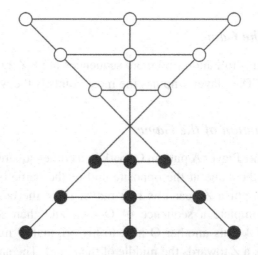

Figure 4.4. Starting position on a lau kata kati game board.

Players alternate turns after deciding who goes first. In turn the player must move into a vacant intersection, or capture an opponent's piece by jumping over it along the game-board lines and landing on a vacant intersection immediately after the jumped piece. The piece jumped is considered "captured" and is removed from the board. Jumping and capturing are mandatory moves. A player may make multiple jumps if there is one empty intersection between the opponent's pieces and a vacant intersection at the end of the jumps.

The Goal of the Game

There are several ways to win at lau kata kati.

(1) If a player cannot make a move or a jump (capture), the game is over and the opponent wins.
(2) If neither player can move or capture, the game is over and the player with more pieces remaining on the board wins. (If both players have the same number of pieces, the game is a draw.)
(3) If a player can capture all of the opponent's pieces, that player is declared the winner.

Sample Simulation of the Game

The first move of the game is forced, since the only empty intersection is at the center. Player A moves one piece into the vacant intersection. Player B now must jump over that piece, removing it from the board. Player A now jumps that piece belonging to Player B and removes it from the board. Play continues in this manner. If no jump is available, the player whose turn it is must move a piece onto a vacant intersection. All moves and jumps must follow the linear-pattern of the game board. Since the first move and capture are forced players should play several times, alternating who goes first.

4.7. Pretwa

The game of pretwa is a two-person strategy game from India. Like many other Indian games of capture, the pieces are placed on a specifically shaped grid, with the only vacant space being the center position. This forces the first move and the first capture to an extent. The game is sometimes referred to as "checkers on a wheel".

Rules of the Game

The game is played on a 19-cell board as shown in Figure 4.5. The board consists of 19 intersections formed by three concentric circles and six radii, each 60° apart as shown. Each player uses nine identifiable chips or checkers – usually black and red chips as in checkers. They are placed on the game board along three consecutive, adjacent radii.

Players decide who goes first. Players alternate turns either moving a piece of their own or capturing an opponent's piece. There are two kinds of moves, referred to as "moves" and "capture."

(1) A "move" consists of moving one piece of their own color to an empty adjacent intersection either along a radius or along a circle.

(2) A "capture" (as in checkers), consists of jumping over an opponent's piece, if the next adjacent space is empty. The jump must be made forwards or backwards along a radius line, or along a circle in a sideways motion. The captured piece is removed from the board.

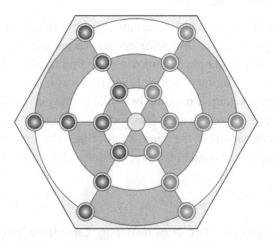

Figure 4.5.   Starting position on a pretwa game board.

(3) A player must make a jump and a capture if one exists. If more than one jump is available, the player whose turn it is can select which of the two available jumps he/she chooses to make.

The Goal of the Game

There are several ways to win at pretwa.

(1) If a player cannot make a move or a jump, the game is over and the opponent wins.
(2) If neither player can move or jump, the game is over and the player with more pieces remaining on the board wins. (If both players have the same number of pieces, the game is a draw.)
(3) The first player who has only three pieces left is the loser.

Sample Simulation of the Game

The first move of the game is forced to some degree. The player whose turn is first must move one of his or her pieces into the center position. This forces the second player to either move along a circle into the space vacated, or jump the piece just moved, and subsequently removing it from the board.

Suppose Player #1 moves along a line, into the center intersection. Now Player #2 must jump from radius into the space vacated by Player #1. The jumped piece is removed from the board. Player #1 now jumps along the originally moved radius, removing a piece from the board. If no jump is available, the player must move a piece along the grid to an available position.

Play continues in this way until one player can no longer move nor jump.

4.8. The Game of Kaooa

The game of kaooa stems from India and mixes capture and blocking strategies. Unlike some of the other capture and block games from India such as pretwa and lau kata kati (see Games 4.6 and 4.7), the board is empty at the start of the game.

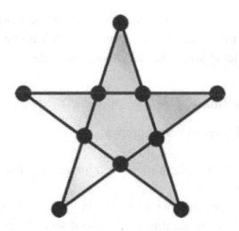

Figure 4.6.   Kaooa game board.

Rules of the Game

The game of Kaooa is played on a pentagram-like board with 10 intersections as shown in Figure 4.6. One player has seven crows; the other player has one raven.

Players decide who will play the seven crows, and who will play the one raven.

There are two kinds of moves, either "dropping" or "moving." The dropping phase takes place at the beginning. The crows go first, after which players alternate turns.

The crow-player drops one crow onto any intersection point on the board. It is now the raven's turn. This player places the raven anywhere on the board. From here on, the crows drop a piece anywhere on the board. The raven can move from one intersection to another adjacent one anywhere on the board in any direction, following the pattern lines shown on the board. If a jump presents itself, the raven can make the jump, capture the crow-piece jumped, and remove it from the board. Once all seven crows are on the board, players move their piece in any direction along lines of the board, one intersection at a time. Crows cannot jump. The raven can jump as in checkers, removing the piece that was jumped over.

The Goal of the Game

The goal of the game for the raven is to remove at least four of the crows so that they can no longer block the raven from moving. The goal of the game for the crows is to block the raven so that it cannot move.

Sample Simulation of the Game

Player A plays the seven crows, player B plays the one raven. Player A places a crow in one of the vertices of the pentagonal board. Player B places the raven in the middle of the board on one of the inner intersections. Player A now drops another crow on another vertex. Player B moves the raven back and forth in the center of the board. It will take seven moves for both players to be ready to move along the board. Once all seven crows are on the board, the raven must try to jump a crow each time he can. Will the raven win? Or can he be pinned against the vertex of the board?

Chapter 5

Miscellaneous Strategy Games

One problem in trying to categorize strategy games is that many people see the games falling into a variety of categories. In this chapter, we will discuss a number of games that, to us, do not fall into any other category, yet continue to exhibit problem-solving strategies.

5.1. Taxman

This game is sometimes referred to as the "Factor Game", since scoring depends on the available factors of each number. A winning strategy could be easily developed after one or two games.

Rules of the Game

A list of the counting numbers from 1 to 50, is arranged as shown in Figure 5.1.

Players alternate turns selecting a number from those left in the array. The number picked is added to that player's score. However, the opposing player – referred to as the Taxman for that round – gets the taxes, which is a score equal to the sum of all the factors of the number picked by the first player and that still remain in the array. Once a number is selected and scored, it is erased from the array. The player who was the Taxman now gets to select a number from the remaining numbers in the array for his or her score. The first player is now the Taxman for this round.

1	2	3	4	5	6	7	8	9	10
11	12	13	14	15	16	17	18	19	20
21	22	23	24	25	26	27	28	29	30
31	32	33	34	35	36	37	38	39	40
41	42	43	44	45	46	47	48	49	50

Figure 5.1.   A starting array for Taxman.

The Goal of the Game

The goal of the game is to have the highest score at the end of the game.

Sample Simulation of the Game

Suppose Player A selects 50. That player's score is thus 50 points, a high score indeed. Now the Taxman, Player B, gets all the factors of 50 left in the array, namely 1, 2, 5, 10, 25 for a total of 43 points. These numbers are then erased from the array. In this example it would now be Player B's turn to select a number, and Player A would be the Taxman, and get all the factors of that number. At this point, the array would now look like that shown in Figure 5.2.

Player B now selects 49. That player's score is now 43 + 49 = 92. However, Player A (the Taxman for this round) gets the factors of 49 that are still in the array. The factors of 49 are 7 and 1. But 1 has already been

		3	4		6	7	8	9	
11	12	13	14	15	16	17	18	19	20
21	22	23	24		26	27	28	29	30
31	32	33	34	35	36	37	38	39	40
41	42	43	44	45	46	47	48	49	

Figure 5.2.

removed, so Player A gets only 7. Thus their respective scores at the end of the first complete round would be:

Player A has 50 + 7 or 57;
Player B has 43 + 49 = 92

It is now Player A's turn. This player might now select 43. That's a prime number, so the Taxman (Player B) gets only 1 which is no longer in the array. So Taxman gets 0.

The scores are now

Player number one has 50 + 7 + 43 = 100;
Player number 2 has 43 + 49 + 0 = 92.

The game continues with the players alternating picks and their opponents getting the factors of that pick that are still in the array. The number and its factors are erased from the array after having been picked and scored. The game ends when all numbers are gone and the array is empty. Are there any numbers which are better choices? Is the largest number always the best choice? Why or why not?

5.2. Game of Squares

This game was first published in 1889 and was called *Dots and Boxes*. It has also been referred to as Pigs in a Pen, Smart Dots, Dot Boxing and so on.

Rules of the Game

Begin with an empty grid of 4 × 4 or 16 dots. Players alternate turns connecting any two adjacent dots horizontally or vertically but not diagonally. A player who completes the four sides of a square gets a point as well as another turn. The player's initial, or some other identifying mark, is placed in the square just completed. The game ends when all lines have been drawn.

Some people prefer to start with a 3 × 3, that is, 9 dot array, as a beginning until they learn the game. Once the smaller size has been mastered then the larger one becomes much more intelligible.

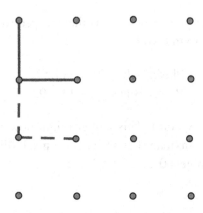

Figure 5.3.   A typical squares game.

The Goal of the Game

The goal is to have more boxes completed at the end of the game than the opponent.

Sample Simulation of the Game

Let's see how a 4 × 4 array plays out. Player A connects two dots along the left side of the array (solid line). Player B connects two dots extending the line (dashed line) from Player A's first move. Player A now connects a dot (solid line) from the middle of this line horizontally. Player B now connects dots (dashed line) the end of the original line horizontally. The board now appears as shown in Figure 5.3 Where should Player A make the next move? Does Player A get a score? Were these good moves? What might have been a better move for Player B?

5.3. Battleship

Commercial versions of this popular game have been available for many years. Computer versions are also available. This is a pencil-and-graph-paper version that appeals to people of all ages.

Rules of the Game

The game is played on a graph-like grid, with the boxes labeled from 1 through 10 both horizontally and vertically as shown in Figure 5.4. Each player has two copies of the grid. On one, the player draws three "ships" each covering three boxes horizontally or vertically, but not diagonally. The player also draws one ship encompassing four boxes. These are drawn anywhere on their own grid. The second blank grid is to keep track of the "shots" the player makes to the setup of the opponent.

Players alternate turn giving the coordinates of what they think is the location of one of their opponent's ships. The opponent must answer either "hit" (if they guessed correctly) or "miss" if they have not hit one of the drawn ships. The shots are specified as a coordinate pair, with the horizontal coordinate written first. For example, in the diagram shown in Figure 5.4, any of the coordinate (3,6), (3,7) or (3,8) would be a "hit", while (5,3) would be a "miss". When the player hits all three (or four) coordinates of a ship, it is considered sunk, and the player must announce it as sunk.

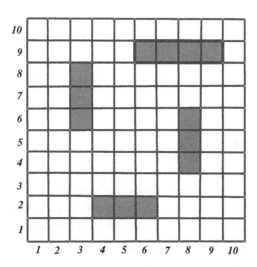

Figure 5.4. A typical starting position for battleship.

The Goal of the Game

The first player to "sink" all of the opponent's ships is the winner.

Sample Simulation of the Game

Suppose the first player selects the box labeled as (1, 1) This is a valid choice, (but not a very good one. Why?) What is a better choice?

The second player number says simply "Miss" and records the guess. It is then the second player's turn; he/she might select (8, 4). The first player, in this case would say "Hit!", since it hit one of the boxes representing a ship.

The first player might now select (4, 2) – also a "hit".

Now both players must decide whether to go to an adjacent cell either horizontally or vertically, and which direction. If the second player chooses (7, 4), he will get a "Miss" response.

Play continues with players alternating turns until one player has all of his or her battle ships "sunk," and is then the loser.

5.4. Solitaire

As the name *Solitaire* indicates, this game is for one player. It matches the skill of the player against the shape of the game board. It can easily be made from a piece of wood using golf tees or matchsticks for pegs in the holes.

Rules of the Game

The game is played on a triangular array as shown in Figure 5.5. Chips or coins are placed in the 14 cells. Only the darkened cell (number 5) is left empty. The cells have been marked for identification purposes only.

A move consists of a piece jumping from a cell over another piece into an empty cell. The piece jumped over is removed from the board leaving another empty cell.

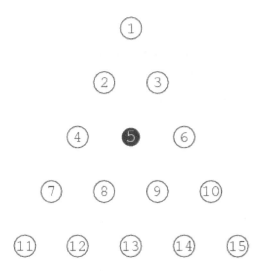

Figure 5.5.   Starting position for solitaire.

The Goal of the Game

The goal of the game is to leave only a single chip on the game board. It's a very difficult task to accomplish, and leaving two, three or even four chips is considered an achievement.

Sample Simulation of the Game

The only empty cell is number five. Thus there are only two possible jumps that can be made. From cell 12, over cell 8 into cell 5, or from cell 14 over cell 9 into cell 5. Suppose we start by jumping from cell 12 into cell 5. The chip in cell number 8 is removed. There are now two empty cells – numbers 8 and 12. There are now a few more options for the next move. For example, we might now jump from cell 10, over cell 9 and into cell 8. We continue in this manner until no more jumps are possible. We then count how many chips are left on the board. The lower the number of chips left on the board when no more moves are possible, the better is the result.

5.5. The Peg Game

This is another strategy game that matches one player the board. It is known by other names as well.

Rules of the Game

The game is played on an 11-cell board. There are five chips of each identifiable color placed as shown in Figure 5.6.

Moves are made according to according to the following rules.

(1) Only one chip can be moved at a time.
(2) A chip can be moved forward if the square it is moving into, is adjacent and vacant.
(3) A chip can jump over a chip of the other color into the next adjacent square, if it is not occupied. A chip cannot jump over a piece of its own color.
(4) No jumped pieces are ever removed from the board.
(5) Chips can only move forward. Moving backwards would involve undoing an earlier move, so the end result would not be a minimum.

The Goal of the Game

The object of this one-player game is to develop a strategy that enables the player to reverse the position of all 10 chips with a *minimum* number of moves.

Sample Simulation of the Game

The player moves his darkened chip into the empty center square. The player has only one choice: he/she must jump with a light colored chip into the empty cell. Remember, a chip cannot jump over a chip of the

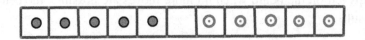

Figure 5.6.   Starting position for the peg game.

same color. The player now has two possible moves; either make a jump move or slide the next color chip forward. If the wrong move is made, it must be undone at a later turn. Thus, the play was no longer a minimum number of moves. The minimum number of moves in this game is 15.

5.6. The Game of Checkers

Throughout this book, we have made reference to the game of Checkers. Many of us have played some form of checkers since our childhood. For those of you who aren't familiar with the game, let's take a quick look at it.

Rules of the Game

Checkers is played on an 8 × 8 square array with squares of alternating colors – dark (usually 32 black) and the other 32 squares lighter (white or red). The game is played entirely on the black squares and all moves are made diagonally. Each player has 12 checkers or chips of his or her own color. Let's assume they are black for Player A and white for Player B. The Players A and B sit opposite each other, and "own" their half of the game board. One player starts the game by moving one of his or her own checkers to a vacant dark square (see Figure 5.7). The row of pieces nearest the player at the end of the board is referred to as the "king" row.

There are two kinds of moves – slide or jump. A slide is moving one box forward diagonally into an empty box. Moves can only go in a forward direction. A jump is leaping diagonally over an opponent's piece into an empty shaded square adjacent to it. The jumped piece is removed from the board. If a jump move is available, it must be taken. Multiple jumps are also possible.

Decide who will go first. After the first move, players alternate turns moving one of their own pieces. All moves are made moving one box at a time, diagonally only, always moving forward. If a player moves one of his or her pieces into the opponent's king row (that is the first row), they are considered a King – designated by placing second checker of the same color on top of it – and can move and jump either forwards or backwards.

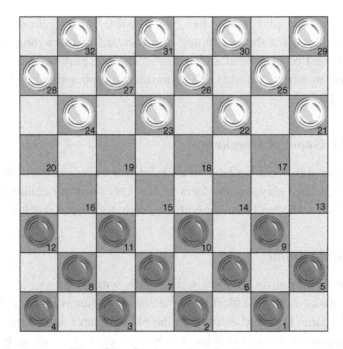

Figure 5.7.   Starting position for checkers.

The Goal of the Game

A player is declared the winner if all the opponent's checkers are removed from the board, or if no move is possible.

Sample Simulation of the Game

The game begins with a single move. Player A moves a piece forward, intending to sacrifice. Player B moves forward into the jump position. If a jump is possible Player A must take it. If not, a regular move forward must be made. The game continues until one player either cannot make a move, or one player has lost all of his or her pieces.

5.7. The Game of 5 × 5

Although this game consists of placing Xs and Os on a grid, in a manner similar to tic-tac-toe, it is not really a version of that game, since the first player to get three-in-a-row does not make that player an immediate winner.

Rules of the Game

The game is played on a 5 × 5 (25 square grid). After it is decided who goes first, each player in turn puts one of his or her marks (X's and O's) anywhere on the game board. Players can decide to block their opponent or play offensively. This game strategy can change throughout the game. After all 25 squares have been selected and marked with O's and X's, the game ends.

The Goal of the Game

Once the game ends, players count the number of runs of three-in-a-row of the same mark. The runs can be vertical, horizontal or diagonal. The player with the most three-cell runs is the winner.

Sample Simulation of the Game

Player A puts an X in the upper corner of the board. Player B puts an O in the middle of the board to try for a horizontal run. Player A decides to block, and puts an X next to the O. Player B changes strategy and goes for a vertical run by putting an O above the first one. Can player A stop Player B? Should Player A move from defense to offense? Was the corner square a good opening move? Why or why not?

5.8. Adugo

The game of Adugo is common among the Bororo tribe of Indians in Brazil. It is similar to many of the hunting games played in Southeast Asia and Africa. Sometimes the players' pieces are known as jaguar and dogs; sometimes they are called tiger and wolves.

Rules of the Game

The game is played on a game board as shown in Figure 5.8. There are 14 "wolves" and one "tiger". The pieces are arranged in a starting position as

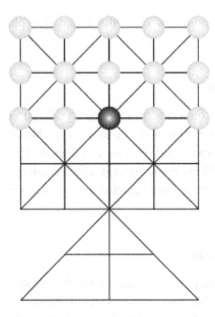

Figure 5.8.   Starting position for Adugo.

shown in the figure. The tiger is hunting the wolves, while the wolves try to trap the tiger.

All moves are made along the lines on the game board onto an empty adjacent line intersection. One player moves the tiger and the other player moves the wolves. The tiger moves first, after which the players alternate turns. Wolves move any one piece in a given turn. Pieces can be moved forwards or backwards. The tiger captures a wolf by jumping over the wolf onto an adjacent empty intersection. The jumped piece is removed from the board. Multiple jumps can be made in a single turn. The wolves cannot jump nor capture the tiger.

The Goal of the Game

The tiger wins if it can capture at least five wolves. The wolves win if they completely surround the tiger, preventing it from moving.

Sample Simulation of the Game

Player A, the tiger, goes first and must move onto any empty intersection. Suppose the tiger moves towards the triangular section of the board. Player B moves a wolf forward into the intersection vacated by the tiger. Player A now moves backwards another intersection, onto the line of the triangle's base. Player B now moves a piece onto the left or right intersection, right behind the first wolf that was moved, creating a double that cannot be jumped. Player B is now forcing the tiger into the triangular section of the Adugo board. Will the tiger be eventually trapped? Or can he jump a wolf and remove it from the board?

5.9. The Game of Awithlaknannai

This is a game often attributed to the Native American Zuni tribe of New Mexico. It goes way back in time; in fact, the exact age of this game is unknown. It is sometimes referred to as Dueling Serpents or Fighting Serpents.

Rules of the Game

The game board consists of a series of isosceles triangle arrangements. Play begins with each player setting up his or her 12 pieces on the game board as shown in Figure 5.9.

Notice that the center intersection remains as the only empty intersection at the start of the game. Players decide who will go first, and from then on, they alternate turns. Players must move along a line on the board.

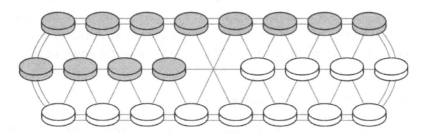

Figure 5.9. Starting position for Awithlaknannai.

There are basically two kinds of moves – capturing or sliding (non-capturing) all along a line on the board. A capturing move is made by jumping over an opponent's piece onto the next empty intersection. The piece that was jumped is removed (or captured) from the board. The turn of the player who made the capture ends. Jumping is mandatory – if a jump exists the player must take it! Multiple jumps with the same piece, if they exist, must also be taken.

If no jump is available, a non-capturing slide must be taken. A slide consists of moving any of the player's pieces along the pattern of the board to an adjacent empty intersection. In this case, no piece is removed from the board.

The Goal of the Game

There are two ways to win at this game. First, the primary goal of the game is to remove all of your opponent's pieces from the board. When this happens, the player with no more pieces on the board loses the game. The second way to win is to block the opponent from making a move – neither a slide nor a jump. If a player has no more moves possible because his or her pieces are all blocked, then that player loses the game. If neither player can move, the winner is the player who has captured more of the opponent's pieces. Otherwise the game is considered a draw.

Sample Simulation of the Game

The first player has no choice of a move. Player A must move a piece into the vacant intersection. Player B must now jump this piece of Player A. Similarly, Player A must now jump Player B's piece. The game continues much like a game of checkers as both players must make a slide or a jump. Is there an advantage to going first? Or would you prefer going second?

5.10. Domino Direction Cover Up

By now, you have probably mastered the game of domino cover up (see Chapter 3, Game 3.3). Domino direction cover up is a slightly different version of that game. The game still requires a traditional 8 × 8 checkerboard and a set of 32 dominoes each of which will cover two squares horizontally or vertically (but not diagonally).

Rules of the Game

The players alternate turns placing a single domino on the board, covering exactly two squares. However, Player A must place dominoes on the board only horizontally, while Player B can only place dominoes vertically. Clearly, no dominoes can be placed diagonally.

The Goal of the Game

A player wins when he/she successfully places a domino on the board and the opponent has no possible way to place a domino anywhere else on the board.

Sample Simulation of the Game

Player A places pieces only horizontally. Player B places pieces only vertically. Player A goes first and places a domino to cover the first two squares on the top row of the game board. Player B now places a piece vertically, blocking the third square on the same row, as well as the square directly underneath. Player A places a second domino to cover squares 4 and 5 in the first row, while Player B now blocks square 6 (see Figure 5.10). Play continues in this same manner. Does going first give an advantage? Who will win? Why?

5.11. Dominoes

The earliest traces of the game of dominoes stem from the thirteenth century in China. Even today, the Eskimos and Inuit Indians of Canada are known to play with sets made from whalebone and containing between 61 and 143 pieces. Today, the classic set of dominoes contains 28 tiles with the largest number of dots being a tile with six on each side. The game can be played by two or three players with this set. There are larger sets of dominoes available to enable more participants in the game, with the

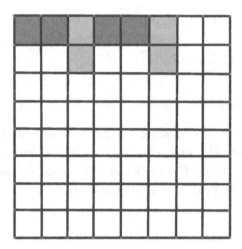

Figure 5.10.   A domino direction cover up game board after two moves.

largest having a tile nine dots on each side. In our discussion, we will consider the classic version which contains 28 tiles.

Rules of the Game

The basic domino set, and the one we are considering in our discussion, contains 28 rectangular tiles, each consisting of a rectangle twice as long as it is wide. Each of the two squares on the face of the rectangle is marked with "dots" or "pips". The reverse side is left blank. The basic set of 28 contains dominoes with dots as shown in Figure 5.11.

There are several versions by which the game of dominoes can be played. For our purposes here we will only consider one variation, where a player who cannot make a required move takes exactly one tile from the pile of unused tiles and forfeits his turn in that round.

We begin the game by placing the 28 dominoes face down and mixing them up, or "shuffling" them by randomly moving them around. Each player now picks seven dominoes and arranges them in a semicircular array in front of him or her. Each of the players can see the faces only of their dominoes; the dots remain hidden from the other player or

Figure 5.11.   The set of 28 dominoes.

players. The remainder of the face down dominoes are referred to as the "pot".

The player with the highest double goes first. (The highest double in the set of 28 dominoes is 6–6, but this need not be the highest domino that begins the game, since that tile may still be in the pot.) If no player has a double, players start picking one domino at a time in sequence from the remaining pile until someone obtains a double which is then placed on the table between the players, face-up.

In turn, players now place one of their dominoes on the table, by matching either end-square's dot number (see Figure 5.12). Players alternate turns in order, placing a domino at either end of any train on the game board by matching the number of dots of the end squares. Here we have shown a tile of double 3–3 as the starting tile. Then a domino of 3–5 touching one end and a 3–0 touching at the other. The next move matched the 0 spot with a 0–2 domino. The next player has a 2–5, and can match at either end.

If a player cannot match, or place, a domino, that player must pick a domino from the pot to add to his or her collection, which ends his turn. (There are some versions of this game where a player continues to

Figure 5.12.   Sample moves to start a domino game.

take tiles from the pot until he or she is able to match one of the end squares. All the while, this player is accumulating a number of tiles that could be a hindrance at the end of the game, since the goal of the game is to get rid of your tiles before the others do.) The player's turn is then over. If a player succeeds in matching one of the ends of a train, that counts as the player's move. Then the next player has his or her turn. This process then continues: at a player's turn, he or she is able to place a tile on the board or if that is not possible, then a tile must be taken from the pot.

A player who can place the last of his or her dominoes on the table "goes out". This player wins the round and gets a score equal to the total number of dots every player has remaining in their collection. If no one can place a tile, the player with the lowest number of dots remaining wins that round and accrues that total number of points in dots remaining in the possession of the other players. At the outset of the game, players should determine what the winning total score should be, or how many rounds should be played before a winner is determined. That is either, as in the first case, when a total number of points is reached, the winner is determined, or, as in the second case, after a certain number of rounds, the player with the highest number of points is determined the winner.

The Goal of the Game

The player who goes out receives the total of all the dots on the remaining player(s) tiles. The winner is the first player to reach a total score decided on in advance or has the most points after a predetermined number of rounds have been played.

(Note: The basic set of dominoes contains only 28 dominoes. This makes the game limited to two or, at most, three players. If more wish to play, use a set of 55 dominoes, one that goes to a double 9 tile. The rules and play of the game would be the same, but the game would last longer.)

Sample Simulation of Game

Here are the first few moves of a sample game with three players using the 28-dommino set. We will assume that player A has the highest double in this game, which is a 5–5, and which he then places on the table. Player B matches one of the 5 spots with a 5–0 domino. Player C matches the other 5 spot with a 5–4 domino. Player A needs a 4 or a 0. If he has neither, he must pick another domino from the pot. Player A's turn is over (see Figure 5.13).

It is then player B's turn. We assume that player B holds a 0–4 domino. Where should this domino be placed? As it now stands, at

Figure 5.13. The beginning of a typical domino game.

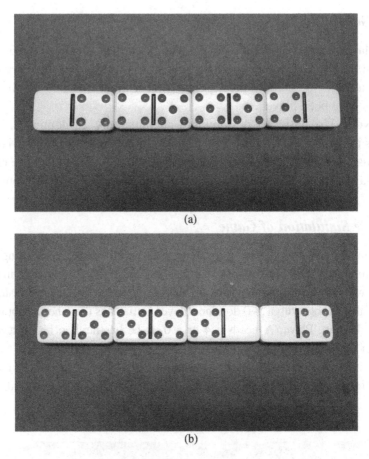

Figure 5.14.   The two possibilities.

whichever end it is placed, the next player must have either a 4 or a 0, since both ends will be the same, either two 4's or two 0's (see Figure 5.14). Play continues until someone goes out or no one can place a domino, at which point the sums of the dots on the remaining tiles of the other players becomes the score for the winner of that round.

Chapter 6

Suggestions and Strategies
for Selected Games

In this chapter, we offer some suggestions and some hints for developing successful strategies for most of the games in this book. In reality, it is most difficult to suggest methods of play that will *guarantee* a win or even a tie. In most cases, a great deal depends on the opponent and his or her level of skill. In many games, there is also a matter of luck. However, what we can do is offer some suggestions that will make you a better player in the various games.

Rather than starting out by reading these "solutions", you should first try to develop your own winning strategies for each game. In some cases, there may be more than one single winning strategy, in the same way that many problems can often be resolved by using more than one problem-solving strategy. Play each game a few times. As you develop a possible strategy, try your technique on another opponent to see if it actually works, or was it just luck?

Chapter 2

2.1. *The Game of Loser's Tic-Tac-Toe*

Once you have mastered the original game of tic-tac-toe and can either win or tie almost every time, several thoughts should come to mind about this particular version of the game. First of all, since there are nine empty cells, one might think it preferable to go second, since this forces your opponent to place five of his marks on the grid, while you need place only four of your marks. However, this doesn't always work. Instead, if you go first, place your mark in one of the side middle cells of the board. These four cells are the ones least likely to get three in a row. If you go second, do the same thing. You will quickly see how this strategy can be effective.

2.2. *The Game of Triangular Tic-Tac-Toe*

The game is very similar to regular tic-tac-toe on the 3 × 3 grid, where three in a row of one player's marks wins the game. However, because of the unusual appearance of the game board, it is important to locate what cells correspond to the center cell (the three inside cells), the side middle cells and outside corner cells. Using the same strategy as in regular tic-tac-toe should prove effective. Here, start by placing your X or O in one of the vertex cells. This should lead to a win or, at the very least, a tie.

2.3. *The Game of ACHI*

The game of ACHI is analogous to regular tic-tac-toe. In fact, placing the six markers on the 9-cell grid is the same as placing the X's and O's onto the nine-cell tic-tac-toe board. The key strategy here is, once again, to try and get your initial three markers into two lines of adjacent cells, trying to set up two potential three-in-a-row winning lines. This makes it impossible for the opponent to block both rows in a single turn. If your opponent has a particular strategy that looks imposing, play defensively and block his or her every move.

2.4. *The Game of Four-in-a-Row*

The objective of scoring four in-a-row includes diagonally. If you look at Figure 6.1 it is the "O" player's turn, where he or she can place their mark anywhere they wish except for the center square as shown with a question mark – that square, if kept open, makes it impossible for "X" to get four in a row diagonally at this move.

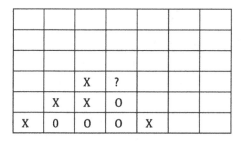

Figure 6.1. A Strategy for a four-in-a-row game board.

Figure 6.2. Bottom row of a four-in-a-row playing surface.

A player should try to force the opponent to move into a vacant square that gives rise to an opening for four in a row in more than one way.

Mathematically, it has been proven that this game is "solved". That is, some mathematicians have shown that if you go first and make the correct moves you must win. If we examine the first or bottom row, and you go first, place your mark in the center box (see Figure 6.2). That is, the box marked as number 4. If you open in boxes 1, 2, 6 or 7, you are almost guaranteed to lose. Boxes 3 and 5 usually result in a draw.

2.5. The Game of Pyramid Tic-Tac-Toe

In this game, the key is to make your first move in the upper single cell. This enables you to begin a potential win in three ways; vertically in the center, or diagonally in either of two ways. On your second move, place your mark in the bottom left (or right) whichever cell is available.

If you are playing a defensive game and you have the second turn, place your mark in the second row end cell. You block one of your opponents potential wins, and start a horizontal win for yourself. This is another game which starts out as a defensive game, and often changes to having one play an offensive strategy. As in many versions of tic-tac-toe, the game often ends in a tie.

2.6. *The Game of Tac-Tix*

Once again, as in most games based on tic-tac-toe, the opening move is important. Since you can win vertically, horizontally or diagonally, you want to establish a position where you have multiple ways to win. Try to get your pieces into a position as shown in Figure 6.3. Notice we can move into cell A-2 or D-2 creating a horizontal win. Or, we can move into C-1, creating a diagonal win. An opponent can only block one of these moves, leaving the other option to enable a win. There are a number of ways to get into a position that will allow three ways to win. This needs to be a strategy to follow throughout.

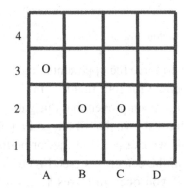

Figure 6.3.   A strategic position for Tac-Tix.

2.7. *The Game of Put Them Down Tic-Tac-Toe*

The player who goes first has an excellent chance of being the winner. Obviously, the center cell is the key to this version of tic-tac-toe. Once a player gets control of the center cell, it becomes comparatively easy to set up two ways to win, making it impossible for the opponent to block both in one move.

An excellent sequence of moves would be to take the upper left corner cell on your first move, and the center cell on the second one. This forces your opponent to take the cell in the lower right to block a diagonal win. Now take the lower left cell and you have two ways to get three in a row.

Once again, a tie is probable if your opponent recognizes your strategy and attempts to block.

2.8. *The Game of Circle Tic-Tac-Toe*

The polar coordinate grid makes this version of tic-tac-toe quite extraordinary. First of all, there is no origin point. Secondly, points are described by the number of a circle and the degrees in an angle. Since the player who goes first has a distinct advantage, the players should decide in advance to play a round of two or four games – alternating who goes first – to decide an ultimate winner.

It is not difficult to recognize four in-a-row on any one circle, or along any one radial segment. The four-in-a-row that is not as easily recognizable is the spiral "line" of four. Look at Figure 6.4. Begin by placing a marker at (1, 330°). At your next turn, place your mark at (2, 0°). If your opponent does not block, go to (4, 60°). Assuming your opponent now blocks, continue along the circle with marks on either side of the previous mark. You can then get three in a row on circle 4, and eventually move towards 4 in a row on circle 4. See Figure 6.4.

If both players are equally skilled, the game may end in a tie as do many versions of tic-tac-toe.

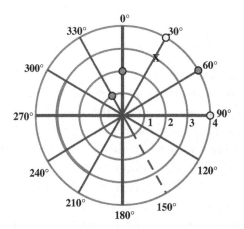

Figure 6.4.   A game board for circle Tic-Tac-Toe.

Chapter 3

3.1. *The Game of Nim*

At the 1940 World's Fair, the Westinghouse Corporation displayed a machine that played nim. This was one of the very first computerized games matching a computer and a human. In this game, we suggest making use of the problem-solving strategy of *working backwards*. That is, thinking from an end position to the start of the game. In order to be certain that you win by picking up the last chip (let's call it number 1), you must leave exactly four chips on the table in your prior move. Then, if your opponent chooses to pick up three chips and leaves one chip, you will win by picking up the last chip. If your opponent selects 2 chips, thus, leaving 2 chips, you will win by picking up both remaining chips. If the opponent picks one, you win by picking up all three remaining chips.

To assure leaving 4 chips on the table, you must also leave 8 chips in the previous move. No matter what your opponent does, you can leave 4 chips when it is your turn. Similarly, to leave 8 chips, make certain you leave 12, 16, 20, 24, and 28 in your respective turns.

To do this, you must go *second*. Whatever your opponent takes, you can leave 28. If you must go *first*, to get to 28 remaining chips, becomes difficult if your opponent knows the strategy.

3.2. *The Game of Reverse Nim*

In this version of nim, you must leave exactly one chip for your opponent to select when it is his or her turn. Thus you must leave your opponent with five chips on your previous turn. If the opponent selects one chip, you choose three chips. If he or she selects two chips, you select two chips also. If three chips are selected, you choose one chip. In each of the cases descried above, only one chip will remain for your opponent to choose, and lose the game. As in the original game of nim, there is a distinct sequence of chips to leave each time your turn comes. This sequence is 29, 25, 21, 17, 13, 9, 5, 1. To do this, you must go first and select three chips leaving 29.

3.3. *The Game of Domino Cover Up*

We need to have 32 dominoes for this game since each domino covers **exactly** two squares. If you place one domino horizontally, in the top row leaving the left-most square open, and then place your next domino vertically in the left-most column just beneath the upper-left square, and thereby blocking that upper-left-most square, you have a good chance to win. Eliminating squares in this manner shortens the game and provides you with an advantage. Again, a tie is possible if both players know the same strategy.

3.4. *The Game of Tromino Cover Up*

To ensure a win, you must go first. Since there are only two shapes of trominos, start by placing a linear tromino to cover the three center squares vertically on the board. Once you have done this, you have ensured a victory. Now, wherever your opponent places a tromino, of either shape you must place your own tromino of the same shape in the exact symmetric (reflection) place on the opposite side of the game board. In this way if your opponent can make a move, so can you! In Figure 6.5, we show two such symmetric moves.

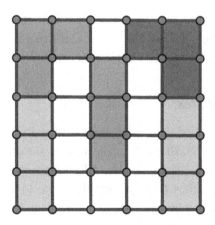

Figure 6.5.

3.5. *The Game of Pathways*

First of all, note that the line segments need not be continuous at the start. You can connect any two adjacent dots anywhere on the game board as long as you do not cross your opponent's path. A path consists of two connected collinear segments. These segments can be connected later. The line to get from one side to the opposite side need not be a straight line. It can twist and turn as it goes across the board, but it must be continuous at the end.

If you go first, begin with a vertical line that connects two central dots on the game board. This provides the opportunity of moving in a wide number of different directions on your next move. The next move could be made horizontally in either direction, and emanating from the top or bottom of the two dots connected on the first move. If your opponent blocks one of the moves, which is established by two connected collinear segments, select the other one on your next turn. Continue extending the pathway by alternating between horizontal and vertical line segments.

If your turn is second, you might consider beginning by blocking your opponent's path nearest to his or her end line. Your strategy will be to force your opponent to one side of the board, thereby, limiting his move options going forward. If you connect two dots horizontally on one side of his end line, you can force your opponent to move along a side of the game board, a very weak position indeed.

As the game progresses, your strategy might change back and forth from offense to defense. Very often you can make a move, which both blocks your opponent and also extends your path.

3.6. *The Game of Hex*

Some people have called the game "John" either for the name of one of its discoverers, John Nash, or because it was sometimes played using hexagonal tiles from the floor of a typical public restroom.

The player who goes first has an advantage to win the game. The game can never end in a tie. For example, a strategy that may be followed is that player X places his or her mark in the cell A, as shown in Figure 6.6. Keep in mind that it is not necessary to make your line continuous at the

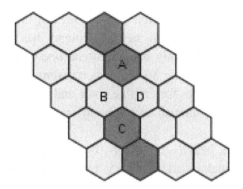

Figure 6.6.

start. Notice that cells B and D are both available to help provide for a continuous path. If your opponent attempts to block your path by selecting cell B, you will have available to you cell D. If your opponent does not attempt to block your path, then move on to cell C, leaving cells B and D for a later move. Continue this way, skipping across the board, leaving various two-adjacent cells empty for later inclusion.

Should your opponent attempt a similar strategy by playing offensive and not defensive, then you, as the first player, will have a distinct advantage towards winning.

3.7. *The Game of Sprouts*

In this game, there is an advantage to going first. In fact, if we start with an odd number of free spots (3, 5, or 7) the first player will win most often. Ironically, if we start with an even number of free spots (such as 4, or 6,) the first player can still win! But a great deal depends on the moves made by your opponent. To best familiarize yourself with the game, begin with three free spots, then gradually increase the number of free spots until you reach seven.

Mathematically, we can show that the number of moves in any game is finite, and someone must win. When we begin, each spot has a potential for three lines in or out, a total of $3n$ potential lines to be drawn, where n is the number of spots we start with. When we connect two spots, we reduce the number lines that can enter these two spots by one each, leaving a total of $3n - 2$ potential lines. However, the new spot added allows for only one new line. This new spot already has one leaving it and one line entering it. Thus, only one additional line can be drawn to that new spot. Therefore, there are $3n - 1$ lines remaining. Each time we add a line we remove two lines for eligible spots and add only one. Thus the number of moves will diminish each time by one until none remain and a player wins.

A win depends heavily on graph theory and topological equivalences, which is mathematical theory too complex to include here. However, if we think of the spot being inside an area (completely enclosed) or outside the enclosed area, you can see that an encircled point has very few opportunities to make a move. Try to avoid these positions. If you see a free spot, then join it to itself. Notice that any spot inside the enclosed area is an – "inside" – point, and cannot be connected to any – "outside" – point. It can only be connected to one of the spots on the perimeter of the loop.

3.8. *The Game of Line-Up-Nim*

At any point in the game there is a mathematical method for deciding if the **current** positions will lead to a win (assuming the player does not make any errors). Mentally, express the number of lines or matchsticks remaining in each row in binary notation, base 2. Thus if we examine the starting position as shown in Section 3.8, the top row contains one line, written as 1. The second row contains two lines, written as 10 in base 2. The third row contains three lines, written as 11, also in base 2. The fourth row contains four lines, written as 100 in base 2. Finally, the bottom row is written as 101, in base 2.

Let's examine the uppermost three rows. Now sum these numbers in columns. For example, if we add the numbers (written in base 2) in rows one, two and three, we get:

Row 1			1
Row 2		1	0
Row 3		1	1

The number of 1s in each column is even – this is a winning position. If there were an odd number, you would have to make changes as you go. Keep the number of **1's** even!

Let's consider an example of actual play. After a few turns, suppose we have three sticks in row 3, two sticks in row 2 and one stick in row 1 as shown above. If player A takes all three sticks in row 3, Player B takes both sticks in row 2, and therefore, Player A, must take the one stick in row 1, and Player A will lose. If player A takes the two sticks in row 2, Player B takes all three sticks from row 3, then Player A must take the one stick from row one and will lose again.

3.9. *The Game of Loop-de-Loop*

One important key to this game is to pay careful attention to the number of points inside and outside of large loops. The key strategy to this game is to leave an odd number of single dots located either inside or outside a larger loop. For example, Figure 6.7 shows a winning position for this game.

Notice that there are exactly three dots left, one inside and two outside the larger left-side loop. It's your turn, so circle one of the two outside dots to itself. There are now two single dots left — one inside and one outside the large loop. Since loops cannot cross one another, your opponent must lose. If your opponent now connects the inside dot to itself, it leaves the single outside dot for you to connect to itself, and the win!

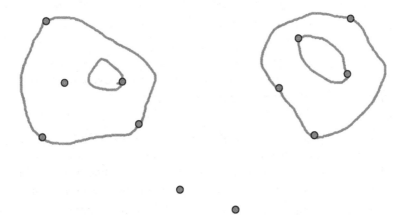

Figure 6.7.   A winning position for loop-de-loop.

Chapter 4

4.1. *The Game of Nine Men's Morris*

Nine Man Mill or Nine Man Morris was always a popular game in Europe. It even was mentioned in William Shakespeare's famous play "A Midsummer Night's Dream".

There are two separate phases to this game. In Phase One, players alternate turns placing their pieces on the board in empty cells. In Phase Two, players move, or slide, their pieces into any adjacent empty cell. Let's refer to Phase One as the "Placing" stage, and Phase Two as the "Moving" stage.

Phase One

The key to winning at Nine Men's Morris is the ability to move into the desirable cells, or spaces. The *corners* are the least desirable places in which to place a chip, since they only have two adjacent spaces to move into in another move. The second square's *intersections* are the best moves, since they have four adjacent spaces into which to move at another turn. *Sides* only have three adjacent spaces, so they are stronger than the corners, but weaker than the intersections. Notice that any piece surrounded by your opponent's pieces is useless. It cannot be moved!

Going first gives a slight advantage. You can place your chips into two adjacent intersections for your first two moves. Try to think ahead! Attempt to place your chips so that you can form more than one row of three. That way if your opponent blocks one row, you can complete the other. Each time you complete a row, you get to remove any one of your opponent's pieces from the game. When you *do* complete one row, remove an opponent's piece that would be most likely to form a row for your opponent on their next move. Remember, you cannot remove an opponent's piece from an already formed mill of three.

The ideal move is to form a row with an empty adjacent cell. Move a piece out of the row on the next move, and then back in on the following move. As long as this movement isn't blocked by your opponent, you can keep forming rows of three and removing pieces of your opponent,

Be careful during the placing stage. Your opponent might build a three man mill, and get to remove one of your pieces. You may have to play defensively if you go second.

4.2. *The Game of Hare and Hounds*

This game, which is sometimes also referred to as the Soldier's Game, is an old game, and after playing the game a few times, strategies begin to emerge, which make it possible for the hounds to win.

In order for the hare to win, he must get behind the hounds, since they cannot move backwards. This necessitates keeping the hare towards the center of the board, and not getting pinned against the edge.

For the hounds to pin the hare, they should keep moving to maintain a straight line moving across the board, and forcing the hare towards a side of the board. Once the hare is on an edge, he can be pinned rather easily.

4.3. *The Game of Kalah*

The player who goes first has a distinct advantage if he or she knows how to play. For example, the best first move for South is to take the three stones in the third pit. They are then deposited in pit #2, pit # 1 and the house! The player gets another turn. This player can now take the four stones in pit #1, and distribute them to his house, opponent's #6, #5, and #4. Notice that South now has two empty pits. If he can end his next turn by moving into empty pit #1 he gets all the stones from his opponent's #6, and deposits them into his house. In each move, the player should try to get at least one stone in his house.

Since there are 36 stones in play, a tie is quite possible. Of course, if both players know the same strategy, a tie game is very likely.

4.4. *The Game of Split-'Em-Up*

Since a pile of two cannot be split, focus on making your opponent leave you with a pile of three. This can only be split into 2 and 1, which should be your final move. Your opponent loses, since there is no remaining valid move option.

As a first move, divide 32 into 29 and 3, leave the pile of three alone. You must strive to always leave an odd number of piles of 3. As an example, if the game reaches a point where only five piles of 3 chips remain, you should select one pile of 3 chips and split it into two and one. Your opponent takes another such pile that leaves three piles of 3 chips. If you take one of these piles of 3, that leaves two remaining piles of 3 chips. Your opponent takes one that leaves one of these piles of 3, and you can then make the last valid move, since the only remaining piles will be a pile of 2 and a pile of 1, and nothing can be done with them.

If you go second, pick up the strategy of dividing into two piles, where one pile has three chips, but leave an even number of these piles whenever possible.

4.5. *The Game of OZO*

To win at OZO, a player should set up as many groupings of four as possible. These groupings should be made as follows:

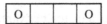

When the opposing player has a turn he/she must place a Z or an O in one of the two open squares. If an O is placed in the second square, you place a Z in the third square, making an OZO. If he/she places a Z in the second square, you place an O in the third square. Again, you have made an OZO to score a point. Since your turn continues, try to make another four-square sequence.

4.6. *Lau Kata Kati*

This game originates from India. As in many of the move and capture games, there is no strategy that will guarantee a win since a great deal depends on the opponent's moves. However, as in the game of checkers, multiple jumps and captures are encouraged. On the defensive side of the game, keeping your pieces in pairs along adjacent lines will force your opponent to move a piece rather than capture one of yours. The first move made by either player **is** forced. Once these moves are made, try to maintain two pieces together.

4.7. *The Game of Pretwa*

The game of pretwa is another one of many games from India played on a fixed grid of circles and lines. The game is similar to checkers except that pieces can go forwards and/or backwards to move or to jump. Like checkers, there is no special strategy, since a great deal depends on your opponent's moves. The best way to avoid losing is to try to keep two of your pieces in an "I" position or if it is three pieces in an "L" configuration, one behind the other or adjacent to one another so no jump is possible by your opponent. If you can keep forcing your opponent backwards, you can eventually keep him/her from making a move.

4.8. *The Game of Kaooa*

The raven must keep away from the vertices of the pentagram, since they pose a direct threat for him to be trapped. It is much easier to trap the raven since his routes for escape are limited. For the crows to win, they must move in blocks of three or four to keep from being jumped or captured. Since it takes the crows seven turns before the player can move any of the pieces, the crows should be dropped on the board in clusters.

Chapter 5

5.1. *The Game of Taxman*

The initial reaction is to choose the largest number on the grid each time when it is your turn. However, this leads to a problem. Suppose Player A selects 50. Then his or her score is 50, a very good score. But how good a score is it? Player B then gets all the factors of 50; that is, 1, 2, 5, 10, and 25 for a total of 43. A better choice would be to consider the odd perfect squares and the primes.

Suppose you select a prime such as 47; now your opponent gets 1 for his score since a prime number only has itself and 1 as factors.

Let's look at the perfect squares. If you select 49 for example, the opponent gets 7 and 1. That's a lot better than 43.

An interesting variation is to play one person against the Taxman. However, in this variation, the player cannot select a number unless *at least* one factor of that number is still in the array for the Taxman to get. Thus, once a prime is chosen and the factor 1 is removed from the array, no other prime number can be selected. Numbers left on the board belong to no one. When no moves remain, add up the scores to see if the player or the Taxman wins.

5.2. *The Game of Squares*

Players who are just beginning to play the Game of Squares usually draw random lines. Here, their only strategy is a negative one. That is, *not* to draw the third side of any square. After playing the game a few times, players usually learn to attempt to build a grouping of one or more adjacent boxes so that any move will give their opponent a chance to take as few squares as possible. At this point in the game, the player who draws the next line tries to take as many squares as possible, and then leave the smallest group of squares for the opponent.

In the example shown in Figure 6.8, we use a basic 3 × 3 dot array. Player B goes second and is using a strategy of symmetry. That is, he/she is copying the move(s) made by Player A to try for a draw!

At move # 7, player A decides to sacrifice a square in the upper-left corner. This gives player B a single square. But, player B must take another turn, and must draw a line somewhere on the board. This gives player A all the remaining squares as shown in move 9. This allows Player A to win.

The key strategy here is to force your opponent to form the first group and take the first square. This concept of sacrificing a single square to obtain several squares in return usually provides a win!

5.3. *The Game of Battleship*

The first few guesses in this game are usually at random. However, once a hit has been found, there are four boxes adjoining this hit. Begin by attacking the box to one side of your "hit". If that is a "miss", try the opposite side. These should be the next few guesses until the orientation of the ship has been found and the ship sunk. Since the players also keep track of misses, any adjacent box that has been recorded as a miss should

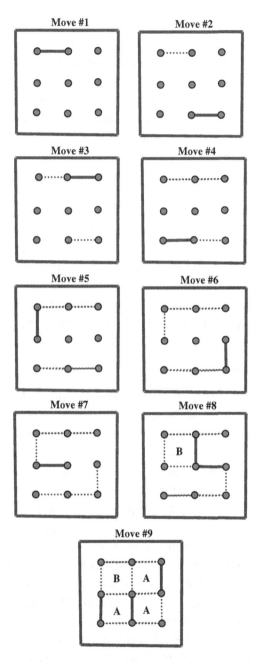

Figure 6.8.   A brief game on a 3 × 3 dot board.

be ignored. Once a second hit has been found, the next cell in either direction should result in a sinking of that ship (for three cell ships).

One defensive strategy is to scatter your ships all around the board. This makes them difficult to find. Other people prefer to keep some of their ships touching either horizontally or vertically or both. This can often confuse your opponent, thereby causing him or her to waste shots while you search for his or her ships.

5.4. *The Game of Solitaire*

The first few jumps in the game are limited by the number and location of the empty cells. The way to develop a successful strategy is to get the chips away from the corners and towards the center of the game board. Once a chip has been isolated in a corner, it can be difficult to remove it from the board. Similarly, the cells along the edges are more difficult to eliminate. Try placing the empty cell in different positions on the board and see if certain placements help in removing more pieces.

5.5. *The PEG Game*

After trying this game a few times players might decide to use the problem-solving strategy **of** finding a pattern from a reduced size and then expanding it to a larger size. That is, reduce the game to one chip of each color and a 3-cell board. Count the minimum number of moves needed and the way the moves are made. Now expand to two chips of each color and a 5-cell game board. Keep track of the *type* of moves made as well as the number of moves. Use "S" to represent a single cell slide, and "J" to represent a jump.

The winning strategy occurs when colors of the chips continually alternate, and no two of the same color are adjacent once they have been moved. Never permit two of the same color pieces to move so that they are adjacent.

The algebraic rule, $M = n(n+2)$, governs the minimum number of moves required. M is the total number of moves, while n is the number of chips of each color. Thus for one chip of each color we need $1 \times 3 = 3$ moves. For two of each color we need $2 \times 4 = 8$ moves. For five chips (as shown), we have $5 \times 7 = 35$ moves altogether.

5.6. *The Game of Checkers*

While there is no real strategy for winning at checkers, there are some suggestions you might follow to help you play better and gain a victory more often. First of all, try to keep your back row (the king's row) as full as possible. That way your opponent cannot get any kings and cannot move backwards. Second, try to keep your pieces on the sides of the board, where they cannot be jumped. At the same time, they force your opponent's pieces towards the middle of the board, where they are more vulnerable. Finally, you may have to sacrifice a single piece in order to jump two of your opponent's pieces. Remember, if a jump presents itself, the player must jump. A sacrifice of one checker that results in jumping two in return is an advantageous move.

5.7. *The Game of* 5 × 5

What makes this game interesting is that concentrating on getting three-in-a-row may be a non-winning strategy. Try to get three in an L-shape, so that two or even three runs of three are possible (horizontally, vertically and diagonally). Don't worry about an opponent getting a single row of three of his or her own marks. Concentrate on setting up multiple blocks of three that can be made into three in a row in more than one way. with just one of your marks. The side squares and corner squares on the board are usually NOT advantageous since one or more possible rows are automatically eliminated by the edge of the board.

5.8. *The Game of Adugo*

This game is known as Komikan in Chile. It is also well known in Asia, which attests to its popularity over the years.

The player who moves the tiger should avoid the intersections in the triangular section of the board. Once trapped in that section, the tiger can no longer escape unless the player with the wolves gets careless. The player who moves the wolves should always attempt to keep them two deep, one behind the other. This way they can avoid being jumped by the tiger. The wolves should be moved slowly across the board staying together as a pack.

5.9. *Awithlaknannai*

The game of Awithlaknannai is much like the game of checkers. That is, pieces either move a single intersection or jump over an opponent's piece to a vacant adjacent intersection, and then the, jumped piece is removed from the board. However, unlike checkers, pieces can move forward or backward as long as they follow the game board lines. Furthermore, much like checkers, games can end in a draw.

The first few moves of the game are dictated by the single empty intersection. The first move must go into the empty intersection in the second row. The second player must jump and capture a piece. Now things change. If a jump is available, it must be made. If not, a slide should be made as far back as possible. As long as players are kept in square blocks of four, no jumps can be made. Try to slide pieces to set up a double jump by sacrificing a single jump and getting the double jump in return.

5.10. *The Game of Domino Direction Cover Up*

There are eight squares in each row and each column of the game board. If Player A goes first and places his or her pieces horizontally, start by placing the first domino so that it will block the first row, second and third squares from the side. On the next move, place the domino so that it will block the second row, the first and second squares from the edge. This not only gives you plenty of room to block your opponent's moves, but also stops your opponent from taking the border squares (see Figure 6.9).

Similarly, if the first player places pieces only vertically, begin by blocking the top row second square from the edge and the second row second square from the edge as shown in Figure 6.10. No horizontal piece can stop you from taking row 1 square 1 and row 2 square 1 later on in the game. Only a vertically placed domino can make use of these squares.

This sequence of moves, if continued in a similar manner, should always leave the player a pair of squares to cover, but, at the same time, will block the opponent. If both players follow this sequence, the game cannot end in a tie. Someone will have to make the last legitimate placement.

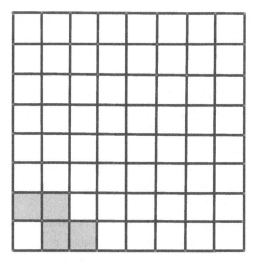

Figure 6.9.

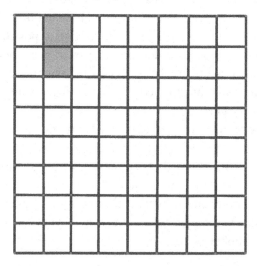

Figure 6.10.

5.11. *The Game of Dominoes*

It would be next to impossible to provide a sure-fire way to win **every** time at dominoes. A great deal of luck is involved. Much depends on the domino tiles you draw from the pot, what dominoes your opponents draw, and what domino he or she plays on the board. However, aside from pure luck, there are some things you *can* do to increase your chances of winning more games than you lose.

First of all, play off your larger numbered dominoes as quickly as you can. If an opponent should unexpectedly win the round, you will at least have tried to minimize the score you must give to that winning player. The more points you have in your collection, the higher your opponent's score might be should they win

Second, if an opponent cannot play a domino at his or her turn, try to remember what number(s) blocked their play so that you can perhaps block them again on a later turn.

A third thing to remember is to try to get rid of your double tiles. While most dominoes provide two different number chances to place them, you have only one opportunity to place a double on the table. Therefore, try to get rid of your double tiles as soon as possible. Furthermore, the point value of double-numbered dominoes can mount up quickly.

Fourth, keep as wide a variety of numbered tiles as possible. Remember, every time you cannot place a domino, you must pick another one from the pot. This increases the number of points in your collection. Should an opponent win the round unexpectedly, you would not want to have a lot of points to offer the winner.

A fifth thing to do is to keep track of what dominoes with a particular value have or have not been played. For example, if many of the dominoes with a five on them have already been played, saving a five-domino is not a strategic move. Keep looking at the dominoes already on the table.

Finally, remember that blocking an opponent causes them to pick another tile. Thus, you may not necessarily be able to block them later on using the same number as before.

Appendix

Game Boards for Various Strategy Games

This chapter will provide you with the game boards where needed. Feel free to copy a game board, enlarge it if necessary, and play the game. Make extra copies of the game board for the game you wish to play. Play it several times to develop a winning strategy.

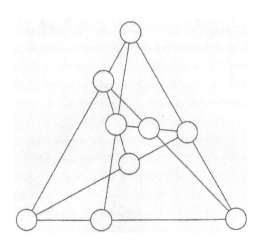

Game 2.2.   Triangular tic-tac-toe.

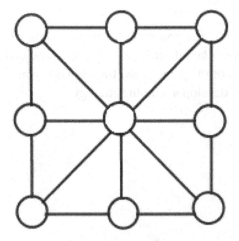

Game 2.3.   Achi.

Game 2.4.   Four-in-a-row.

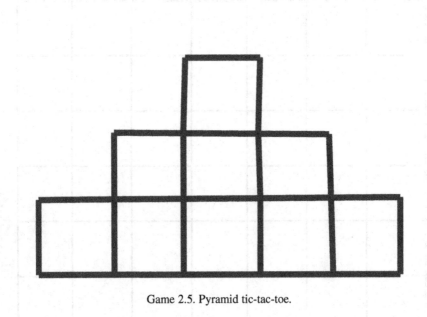

Game 2.5. Pyramid tic-tac-toe.

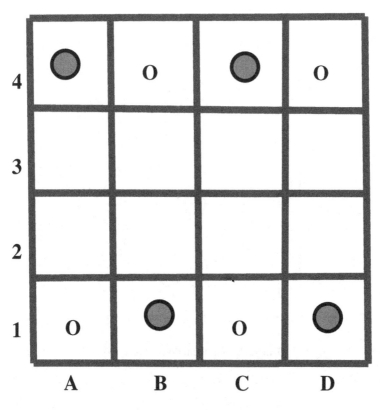

Game 2.6.   Tac-tix.

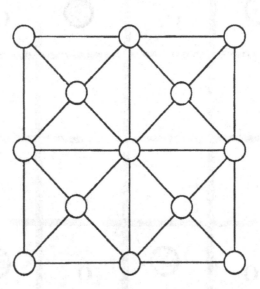

Game 2.7.   Put them down tic-tac-toe.

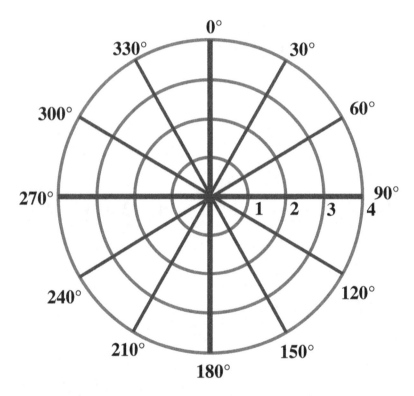

Game 2.8.   Circle tic-tac-toe.

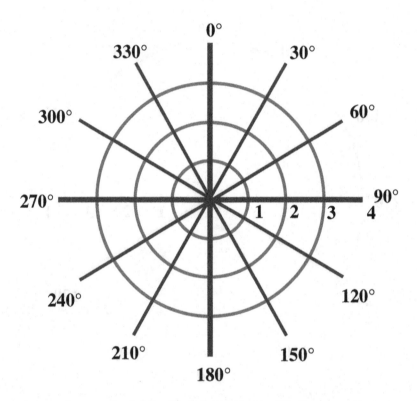

Game 2.9.   Circle tic-tac-toe (simplified).

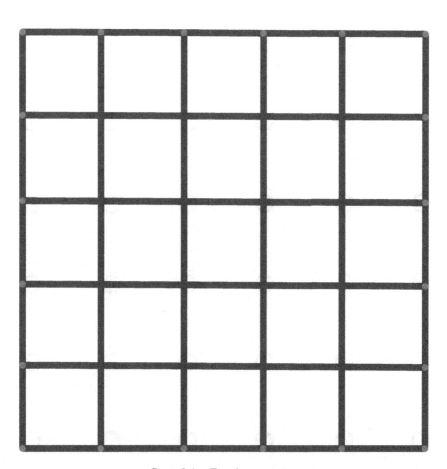

Game 3.4.   Tromino cover up.

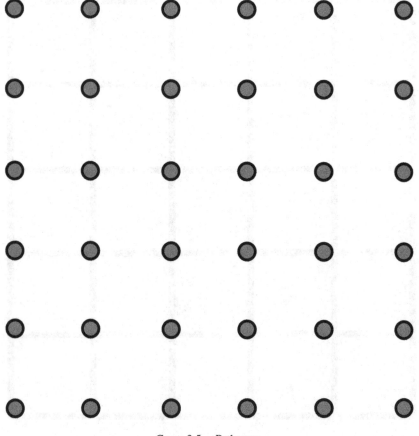

Game 3.5.   Pathways.

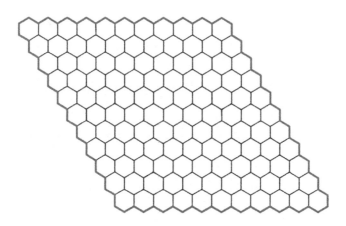

Game 3.6.   Hex.

Game 3.7.   Sprouts.

Game 3.9. Loop-de-loop.

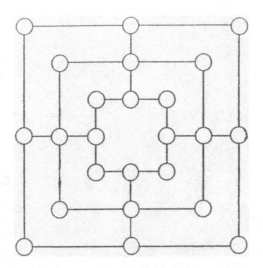

Game 4.1.   Nine-Man's-Morris.

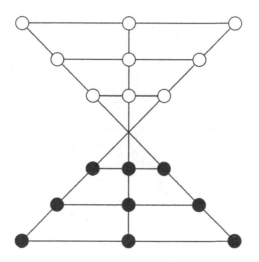

Game 4.6.   Lau kata kati.

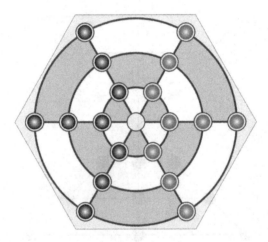

Game 4.7.   Pretwa.

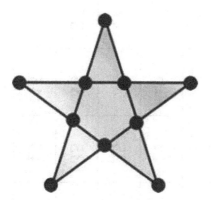

Game 4.8.   Kaooa.

Problem-Solving Ability in Mathematics

1	2	3	4	5	6	7	8	9	10
11	12	13	14	15	16	17	18	19	20
21	22	23	24	25	26	27	28	29	30
31	32	33	34	35	36	37	38	39	40
41	42	43	44	45	46	47	48	49	50

Game 5.1.   Taxman.

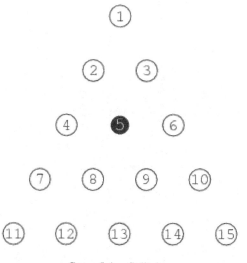

Game 5.4.   Solitaire.

Game 5.5.   Peg game.

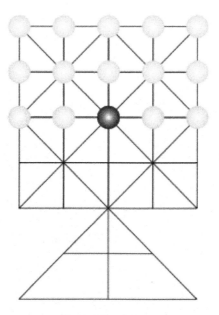

Game 5.8.   Adugo.

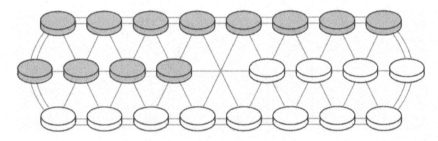

Game 5.9.   Awithlaknannai.

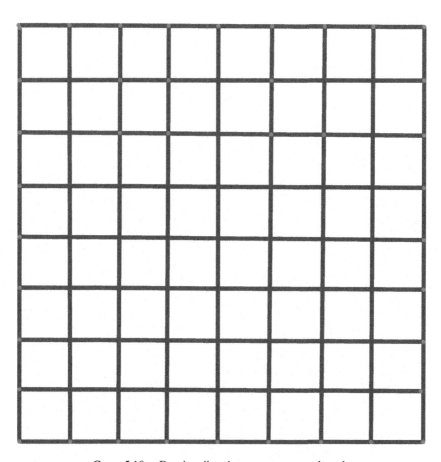

Game 5.10.   Domino direction cover-up game board.

Index

A

Achi, game of, 11–12
  game board, 94
  goal, 12
  origin, 11
  rules, 11–12
  sample simulation, 12
  suggestions and strategies for, 72
Adugo, game of
  game board, 113
  goal, 62
  rules, 61–62
  sample simulation, 63
  starting position for, 62
  suggestions and strategies for, 89
  wolves and tigers, 61–62
Al-Qurna, Egypt, 11
Awithlaknannai, game of, 63
  game board, 114
  goal, 64
  origin, 63
  rules, 63–64
  sample simulation, 64
  starting position for, 63
  suggestions and strategies for, 90

B

Battleship, game of
  goal, 56
  hit and miss, 55–56
  rules, 55
  sample simulation, 56
  suggestions and strategies for,
    86–88
  typical starting position, 55
Blocking games, 23–36

C

Champion, Jr., William Julius, 40
Checkers, game of, 59
  goal, 60
  rules, 59
  sample simulation, 60
  starting position for, 60
  suggestions and strategies for, 89
Checkers on a wheel. *see* Pretwa
Circle tic-tac-toe, game of, 18
  game board, 18, 99
  goal, 19
  rules, 18–19
  sample circle, 20

sample simulation, 19
simpler circle, 21
simplified, 20
game board, 21, 100
goal, 20
rules, 20
sample simulation, 21
suggestions and strategies for, 75
wining positions, 19
Confucius, 4
Conway, John H., 28
Crows and ravens. *see* Pretwa

D
Domino cover up, game of, 25
goal, 25
rules, 25
sample simulation, 25
suggestions and strategies for, 77
Domino direction cover-up, game of, 64–65
blocking opponent, 91
game board, 115
goal, 65
horizontal vs vertically placed domino, 90–91
rules, 65
sample simulation, 65
suggestions and strategies for, 90
winning strategy, 91
Dominoes, game of, 65–66
beginning of typical domino game, 69
domino direction cover up game board after two moves, 66
dots and pips, 66
goal, 69
horizontal vs vertically placed domino, 91–92
origin, 65

"pot," 67
rules, 66–68
sample moves to start a domino game, 68
sample simulation, 69–70
set of 28, 67
two possibilities, 70
Dot boxing. *see* squares, game of
Dots and boxes. *see* squares, game of

F
Factor game. *see* taxman, game of
5 × 5, game of, 60–61
goal, 61
rules, 61
sample simulation, 61
suggestions and strategies for, 89
Four-in-a-row, game of, 13
bottom row of four-in-a-row playing surface, 73
game board, 95
goal, 13
rules, 12–13
sample simulation, 13
sequence of moves in four-in-a-row, 14
strategy for, 73
suggestions and strategies for, 72–73

G
Games
requiring ongoing change of strategies, 37–49
suggestions and strategies for selected, 71

H
Hare and hounds, game of
goal, 39

rules, 39
sample simulation, 39
suggestions and strategies for, 83
Hein, Piet, 28
Hex, game of, 28
game board, 29, 103
goal, 29
rules, 28
sample simulation, 29
suggestions and strategies for,
78–79

K

Kalah, game of, 40
game board in starting position,
40
goal, 41
rules, 40–41
sample simulation, 41
suggestions and strategies for, 83
Kaooa, game of, 47
board, 48
"dropping" or "moving," 48
game board, 109
rules of game, 48
sample simulation of game, 49
suggestions and strategies for, 85
Krulik, Stephen, xii

L

Lau kata kati, game of, 44
game board, 107
goal, 45
rules, 44–45
sample simulation, 45
starting position on lau kata kati
game board, 44
suggestions and strategies for, 84
Line-up-nim, game of, 33
goal, 34

rules, 33
sample moves, 34
sample simulation of, 34–35
starting position for, 34
suggestions and strategies for,
80–81
Loop-de-loop, game of, 35
first three moves, 36
game board, 105
goal, 36
rules, 35
sample simulation, 36
suggestions and strategies for,
81
typical array of dots for, 35
winning position for, 81
Loser's tic-tac-toe, game of, 23
goal, 24
rules, 23
sample simulation, 10
suggestions and strategies for,
71–72

M

Miscellaneous strategy games, 51–70

N

Nash, John, 28
Nim, game of, 23
goal of, 23
rules of, 23
sample simulation of, 24
suggestions and strategies for,
76
Nine Men's Morris, game of, 37
game board, 106
goal of game, 38
origin, 37
phase one, 82–83
rules of game, 37–38

sample simulation of game,
    38–39
suggestions and strategies for, 82
Noughts and Crosses. *see* Tic-tac-toe
    board

O
OZO, 42
    goal of game, 43
    rules of game, 42–43
    sample simulation of game,
        43–44
    suggestions and strategies for, 84

P
Paterson, Michael S., 29
Pathways, game of, 27
    game board, 27, 102
    goal of, 28
    rules of, 27–28
    sample simulation of, 28
    suggestions and strategies for, 78
Peg, game of, 58
    game board, 112
    goal of, 58
    rules of, 58
    sample simulation, 58–59
    starting position for, 58
    suggestions and strategies for, 88
Pigs in a Pen. *see* squares, game of
Posamentier, Alfred S., ix
Pretwa, 45–46
    game board, 108
    goal of, 47
    "moves" and "capture," 46
    rules of, 46–47
    sample simulation of, 47
    starting position on pretwa game
        board, 46

suggestions and strategies for, 85
Problem solving
    heuristic skill, 2
    vs. strategy games winning, 2–3
Put them down tic-tac-toe, game of,
    17
    game board, 98
    goal of, 16
    rules of, 16
    sample simulation of, 17–18
    suggestions and strategies for,
        74–75
Pyramid tic-tac-toe, game of, 14
    game board for pyramid
        tic-tac-toe, 29
    goal of, 14
    rules of, 14
    sample simulation of, 14

R
Resolve problem situations, ability
    to, 1
Reverse nim, game of, 24
    goals, 24
    rules, 24
    sample simulation, 24
    suggestions and strategies for, 76

S
Smart Dots. *see* squares, game of
Solitaire, game of, 56
    game board, 111
    goal, 57
    rules, 56
    sample simulation, 57
    starting position for, 57
    suggestions and strategies for, 88
Split-'em-up, game of, 41
    goal, 42

rules, 42
sample simulation, 42
suggestions and strategies for,
   83–84
Sprouts, game of, 29
game board, 104
goal, 31
possible moves in, 30
rules, 30–31
sample simulation, 31–33
starting, 30
suggestions and strategies for,
   79–80
Squares, game of, 53
brief game on a 3 × 3 dot board,
   87
goal of, 54
rules of, 53–54
sample simulation, 54
suggestions and strategies for, 86
Strategy games
finding and creating, 3
problem solving *vs.,* 2–3
time management, 2
winning, game of, 2

T

Tac-tix, game of
game board, 97
goal of, 15
rules of, 15
sample simulation of, 16
starting position for, 16
strategic position for, 74
suggestions and strategies for, 74
Taxman, game of, 51
game board, 110

goal, 52
rules, 51
sample simulation, 52–53
starting array for, 52
suggestions and strategies for,
   85–86
Tic-tac-toe, game of, 9–21. *see also*
   triangular tic-tac-toe
board, 4
change in game board, 6
change in rule to winning, 5
goal, 10
losing, 9
opportunity to utilize
   mathematical skills, 6
rules, 9
sample simulation, 10
strategy to winning in, 4–5
Time management, strategy games, 2
Triangular tic-tac-toe, game of, 10
game board, 10, 93
goal, 11
rules, 11
sample simulation, 11
suggestions and strategies for,
   72
Tromino cover up, game of, 25
game board, 26, 101
goal, 26
rules, 26
sample simulation, 27
suggestions and strategies for, 77
tromino shapes, 26
Two-person strategy game, 2

Y

Yih. see Tic-tac-toe board

Printed in the United States
By Bookmasters